CLEANING YOURSELF TO DEATH
How Safe is Your Home?

CLEANING YOURSELF TO DEATH
How Safe is Your Home?

PAT THOMAS

Newleaf

Newleaf
an imprint of
Gill & Macmillan Ltd
Hume Avenue, Park West, Dublin 12
with associated companies throughout the world
www.gillmacmillan.ie
© Pat Thomas 2001
0 7171 3162 9
Index compiled by Cover To Cover
Print origination by
Carrigboy Typesetting Services, Co. Cork
Printed by ColourBooks Ltd, Dublin

This book is typeset in 10pt on 13pt Galliard.

*The paper used in this book comes from the wood pulp of managed forests.
For every tree felled, at least one tree is planted, thereby renewing
natural resources.*

A CIP catalogue record for this book is available from the
British Library.

1 3 5 4 2

Contents

Foreword   vii

Acknowledgements   xi

Part 1 Putting Yourself in the Picture   1
    1. Pollution Begins at Home   3
    2. Getting Beyond the Hype   16
    3. Wake up and Smell the Chemicals   29
    4. How Your Body Reacts   39

Part 2 Toxic Toiletries   59
    5. Bath Soaps and Body Washes   61
    6. Dental Care   71
    7. Deodorants and Antiperspirants   83
    8. Facial Care   93
    9. Hair Care   106
   10. Skin Care   127

Part 3 Harmful Household Products   137
   11. Air Fresheners   139
   12. All-Purpose Cleaners   146
   13. Bathroom Cleaners   153
   14. Carpet and Upholstery Cleaners   161
   15. Dish Detergents   167
   16. Floor Cleaners   173
   17. Glass and Window Cleaners   178
   18. Laundry Products and Fabric Care   182
   19. Oven Cleaners and Drain Cleaners   193
   20. Polishes   196

Postscript   205

Glossary of Terms   207

Useful Contacts   211

Select Bibliography   216

Index   222

Foreword

Most books about toxic chemicals are about saving the planet. This is a book about saving ourselves.

When I was at school, people who liked science and chemistry were nerds. The kids who could pronounce those long chemical names and who could get their chemistry experiments to work properly, or whose science projects were actually scientific, seemed to inhabit a different world from the rest of us and pretty much existed to be teased and poked fun at.

Today it turns out the joke is on the rest of us because chemicals rule our lives. They are in everything we eat and drink and every lungful of air we take in. They are on the clothes we wear, on the furniture we buy and in the cars we drive. And the emerging evidence is that many of these chemicals have the potential to make us very ill.

If you look at a typical bottle of shampoo, you may see as many as 20 different ingredients listed on the label, the majority of which have been synthesised in the lab. To the average person the list of ingredients is just an incomprehensible alphabet soup. Words with more than three syllables, names beginning with hydroxy-this and poly-that simply make most people switch off. Instead we just cross our fingers and pray that the companies who have put this chemical cocktail together really do have the consumer's welfare and best interests at heart.

As the writing for this book progressed it became clear that faith and prayer were not enough to guarantee the safety of the products we use every day. Instead, information is the key. Only by allowing your hidden nerd to emerge and by getting into the habit of reading the labels on the products you use everyday can you begin to make intelligent and safe choices of the products you use. By refusing to use products that are toxic, by considering safer, even home-made, alternatives you are sending out a strong message to manufacturers that the wanton chemical poisoning of our bodies, and our planet, is simply not acceptable.

If it all sounds like too much hard work, consider that it was not so long ago that we all had to learn another new language – computer language. We had to learn our bits from our bytes and get to grips with ISPs and htmls. Many of us were dragged kicking and screaming into the computer age, but once there found that it revolutionised our lives.

Some of the information in this book is shocking. Consider the following, for example:

- The typical home contains an average of 63 hazardous products that together contain hundreds of different chemicals.

- In the US, the National Institute of Occupational Safety and Health analysed 2,983 chemicals found in personal care products – 884 (30 per cent) were deemed toxic.

- A report from the European Chemicals Bureau concluded that only 14 per cent of chemicals produced in or imported into the EU have the most basic set of safety data publicly available.

- Women who work at home have a 54 per cent higher death rate from cancer than those who work outside of the home.

- When scientists from the US Environmental Protection Agency (EPA) tested a random selection of 7,000 people they found toxic chemicals in the urine of 71 per cent of them.

- The EPA has also found that the air quality in homes is much more toxic than the outdoor air – often containing between 2 and 5 times the concentration of toxic chemicals.

- You can absorb up to 60 per cent of any substance applied to your skin.

Taking these facts on board has important implications for our health and for the health of future generations. This book provides a starting point, with information on which chemicals are most dangerous, where you are likely to encounter them, and how they behave in your body. It also examines the intense commercial pressure consumers are under to keep buying and using potentially dangerous products as well as ways of breaking our dependence on them. It is not always happy reading, but I profoundly hope it is helpful reading.

Pat Thomas
London 2001

Acknowledgements

My thanks go out once again to my family and friends who were patient and supportive while I researched and wrote this book.

I am indebted to all at Gill & Macmillan, especially Eveleen Coyle who understood and enthusiastically supported the writing of this book from the word go. Also to Laura Longrigg who continues to be more than just an agent.

I am also very grateful to the authors, environmentalists, toxicologists and scientists, whose work is listed in the bibliography, for all they have done in bringing the vital issue of toxic chemicals to the fore.

Finally thanks to Lynne McTaggart and Bryan Hubbard at What Doctors Don't Tell You for publishing my article 'Toxic Toiletries' in 1999 which proved to be the springboard for this book.

Good health and love to you all.

Part 1

Putting Yourself in the Picture

Chapter 1

Pollution Begins at Home

Cleanliness is next to godliness . . . or so they say.

Most of us tend to think of our homes as havens, and environmental pollution as a problem that is 'out there'. Perhaps there was a time when this was true. But not any more. Today some of the most toxic chemicals we come into contact with are not blown in through the window from some anonymous factory or a passing car. They are bought in good faith in stores and supermarkets and brought back into our homes by us.

It may sound unbelievable but many scientists now recognise that air inside the home (as well as in offices and cars) may be more seriously polluted than outdoor air, even in the largest and most industrialised cities. And since we spend as much as 90 per cent of our time indoors, the health risks from indoor pollution are much greater than those from outdoor pollution.

There are many sources of indoor pollution. Some have received more publicity than others. But one aspect of indoor pollution that has been largely ignored is that which comes from everyday cleaning products. Because of their association with good hygiene, personal and household cleaning agents are the most deceptive of indoor pollutants. While many people are aware that these products can contain allergens, few are aware that they can contain other substances that have the potential to affect human health in other, more substantial, ways.

Such cleaning agents can, for instance, contain carcinogens (cancer-causing substances), hormone-disrupting chemicals, central nervous system disrupters, reproductive toxins and psychoactive chemicals (substances that alter brain function). The awful truth, in many cases, is that when we rub these noxious chemicals into our skin, or inhale them during steamy showers, when we rinse our mouths, or wash our clothes, dishes and floors with them, we are using hazardous waste to wash away simple dirt.

That may sound over-stated. However, as you read through this book you will see that many of the products we use every day contain chemicals officially designated as 'hazardous waste' by government agencies. It is hardly surprising, then, that there is now accumulating evidence to suggest that by using these products we may be washing ourselves into a state of chronic ill-health and perhaps even an early grave.

The Bigger Picture

When it comes to toxic exposure in the home, the old saying 'what you don't know can't hurt you' is not always good advice. This book is written from the opposite standpoint – what you *do* know can help you. While it is focused exclusively on products used to clean our bodies and our homes, such products are only one element in the bigger picture of how our bodies become polluted.

Today we ingest dozens of harmful chemicals when we eat conventionally grown produce; fertilisers, herbicides, pesticides and fungicides all combine to make so-called 'fresh' food a significant source of poisons. When we eat conventionally reared meats we are ingesting growth hormones and the myriad of just-in-case medications given to conventionally farmed animals, in addition to the pesticides, herbicides, fertilisers and fungicides contained in their feed. As the BSE, or 'mad cow', saga has shown us we are also sometimes ingesting deadly micro-organisms.

Brightly coloured plastics in our homes may look modern and clean, but they continually give off toxic formaldehyde gas which is implicated in respiratory problems and has also been shown to be carcinogenic. Formaldehyde also comes from soft furnishings, vinyl wallpapers, insulation, varnishes, new carpets, upholstery and new mattresses.

Ever had a good whiff of a new book or newspaper? If so you were inhaling toxic gases from the paper processing and the ink. Thirsty? Tap water has become a significant source of many different toxins including pesticides, hormones, hazardous waste and heavy metals, as well as water-borne parasites. If you live in an area where the water is fluoridated your water contains a chemical that is more toxic than arsenic and only slightly less toxic than lead.

In addition to chemical pollution, we are also subjected to electronic pollution. As our offices and homes become more crammed with 'necessary' equipment, we receive a daily onslaught of electromagnetic fields (EMFs). The link between EMFs and cancer as well as other chronic conditions such as headaches, fatigue and mental confusion is becoming firmly established with each passing year.

Given all this you may wonder how we manage to stay healthy at all. The reality is we don't. Even though we are now living longer than ever, we are strikingly unhealthy. Chronic diseases are on the rise. Respiratory problems such as asthma and bronchitis have doubled in recent years, with the young being most affected. Vague disorders such as sinusitis and allergic rhinitis are becoming major problems, especially in inner-city areas. Heart disease, diabetes and thyroid problems are on the rise too. Infertility, both male and female, is becoming more common as are other hormonally linked disorders. Cancer continues to be a persistent disease across all cultures, age groups and genders, with no medical solution in sight.

This is not a very encouraging picture. But what is worse is that we have allowed ourselves believe that such conditions are a normal part of being human and a normal part of ageing.

There is nothing normal about disease. The human body is a robust and remarkable organism which is oriented towards survival and wellness. Without our even being aware of it our bodies work daily to maintain balance and eliminate the toxic by-products of modern life. But even this most perfect system cannot function indefinitely without some help from its owner.

Disease occurs because our natural defence systems have broken down or because our normally reliable organs such as heart, liver and kidneys have been so mistreated they cannot function effectively any more. Some forms of pollution are unavoidable – that is the tragedy and the risk of modern life. But many more forms of pollution are completely within our control. The more information you have about the products you use, the better equipped you are to begin the process of weighing up the risks and benefits of each.

Taking Risks

Weighing up risks and benefits is never easy. As our society becomes more complex, so does the job of assessing our own safety. Making a decision about risk is also particularly difficult when basic information is lacking. At the moment the average consumer has only years of habit and the constant background of TV, radio and magazine advertisements to provide information about the products we use.

Manufacturers of toiletries are required to disclose the contents of a product on the label. However, without some means of decoding the chemical names, most consumers would not be able to tell which products to avoid and which ones are relatively safe. Worse, manufacturers of household cleaning products are not obliged to disclose their ingredients at all. Instead they are allowed to describe them (e.g. 'non-ionic surfactant'). This means that most consumers can only guess at what these products contain.

In addition, most consumers are not well informed about the difference between 'probability' and 'risk'. When manufacturers say that the chances of you getting cancer from a particular ingredient are one in a million they are talking about probability. Calculating probability is a number-crunching exercise, the results of which can be terribly misleading.

When you see that one in a million figure, it is a projection based on averages. It does not take into account variables such as lifestyle and environment. Nor does it take into account the fact that few of us actually resemble the fictitious 'average person' on whom estimates of probability are based – usually a 12½ stone (180 pound) white male. Thus, in any group of a million people exposed to the same chemical none may get cancer – or 100 may get cancer.

What has not been determined, what no company has *ever* determined, is what happens when Mr or Ms Average begins using lots of other products, with lots of other chemicals, in combination with each other on a daily basis. This is a risk that can never be worked into a probability figure, but one that individual users should be aware of.

Risk is something particularly difficult to quantify. Every day we all take different kinds of risks. Each of these must be individually weighed up against the potential benefits we gain from a particular activity. This is why it can also be misleading to try and compare the risk of one type of activity to another.

For instance, the risks and benefits of driving a car are not the same as those of using a particular shampoo or fluoride toothpaste. In driving the car to the supermarket you may risk a fatal crash, you may risk losing your licence due to careless driving, or you may risk getting an expensive parking ticket. But you may also benefit from not having to lug groceries home on a crowded bus. It may save you time and energy, both of which may be in short supply. Also you may have no other viable alternative to using the car, which means that this particular choice is not really a 'choice' at all.

Assessing the individual risks and benefits of personal care and cleaning products is rather more complex because no two people will respond to the same toxic exposures in exactly the same way. Our emotional, physical and biochemical make-ups, our lifestyles, daily habits, living and working environments all combine to contribute to our unique responses to toxins.

Also, you may like to have a nice clean home, but if you have a family history of cancer, should you risk using a product that contains carcinogens? Equally a bath foam or bubble bath may produce lots of fun and sensuous bubbles, but if it raises your risk of chronic urinary tract infections or leaves your skin dry and irritated, is the benefit of the bubbles worth the risk of physical harm? You may like the fact that a certain toothpaste cleans your teeth and gives you a minty fresh jolt every morning. But if there is enough fluoride in the tube to kill your 5-year-old child, or if the harsh detergent in the toothpaste can cause unsightly mouth ulcers, you may reason that, given good alternatives, the risk is not worth the benefit.

Asking Questions

Whenever challenging information like this comes to light it is reasonable to have lots of questions. Parts 2 and 3 of this book address specific questions about everyday products. Below are the answers to some common general questions about toxins in the home.

- *There must be some laws protecting the consumer. Surely they wouldn't allow toxic chemicals in products intended for use in the home?*

There is no 'they' protecting the consumer. If you read the newspapers you will be aware that literally hundreds of products are pulled off the shelf every year because (and usually after) they have been found to be a danger to the public. Years ago the drug thalidamide was considered perfectly safe to use during pregnancy. Today we have hundreds of people deprived of limbs because their mothers took this 'perfectly safe' drug.

Every year cars are recalled to the factory because of dangerous malfunctions that put our lives at risk. British beef (and increasingly beef from other countries) is another fine example of the way officials try to convince the public that something is perfectly safe – until it is proven to be harmful.

Lots of research has been done into the environmental effects of toxic chemicals and jumping on the environmental or 'green' bandwagon has become a fashionable thing to do. But while we have been busy saving the earth, we have neglected to look after ourselves. No one has ever fully researched the extent to which people come into direct contact with pollutants. Scientists will tell you that it is too complicated, that there are too many variables, and that no one can get proper funding for such an ambitious project. This lack of data is cavalierly taken as proof that certain chemicals are safe.

Neither government nor green agencies can state with any certainty the number of people affected by toxic exposures, the severity of those exposures, or the specific sources of worrisome chemicals. Because of this official bodies have given more attention to the obvious sources of pollution and have maintained their focus on environmental emissions rather than human exposure. Worthy as this is, it overlooks an important point – toxic chemicals only produce health problems when they come into contact with the human body.

In 1980 researchers from the US Environmental Protection Agency (EPA) changed what we know about toxic chemical exposure. Results of their large-scale, multi-centre investigation showed that people were most likely to have the greatest contact with toxic substances within places where there were no official regulations on the types or amounts of chemicals used, places they usually consider to be safe such as homes, offices and automobiles.

Today there are still no regulations on the chemicals used in these 'safe' places and it would seem foolhardy to wait until some official body somewhere decides to implement them. The only person who can protect you and your family from these potentially harmful chemicals is *you*.

• *Maybe household cleaning products do contain dangerous chemicals, but I cannot believe toiletries do as well.*

Certainly there is nothing in the daily barrage of commercials, magazine ads and billboard images to suggest that soaping and scrubbing and moisturising is anything but beneficial to our health.

Yet, underneath the glamorous image presented by advertising there is a less appealing truth. Most of the chemicals that go into our toiletries are no different from the harsh, toxic chemicals used in industry.

For example, propylene glycol (PG) is a wetting agent and solvent used in make-up, hair care products, deodorants and aftershave. It is also the main ingredient in anti-freeze and brake fluid.

Similarly, a related agent, polyethylene glycol (PEG), found in most skin cleansers, is a powerful solvent sometimes found in oven cleaners.

Isopropanol is an alcohol used in hair colour rinses, hand lotions, hair sprays and fragrances. It is a neurotoxic solvent also found in shellac.

Sodium lauryl sulphate (SLS), used in toothpastes, shampoos and just about every personal cleaning solution, is a harsh detergent commonly used as an engine degreaser.

Each of these ingredients readily penetrates the skin and has adverse consequences for our health. Some are irritants, some are allergenic, some can damage internal organs and some are carcinogenic. Far from enhancing health, exposure to such chemicals poses a daily threat to it.

However, the most dangerous chemicals that we put in and on our bodies in the name of hygiene and beauty each day belong to a family of hormone-disrupting chemicals that are water-soluble ammonia derivatives. DEA (diethanolamine) is almost always contained in products that foam including bubble bath, body washes, shampoos, soaps and facial cleansers. It is used to thicken, wet, alkalise and act as a detergent. While causing irritation to the skin, eyes and respiratory tract, DEA and its related compounds such as TEA (triethanolamine) and MEA (monoethanolamine) are not con-sidered carcinogenic in themselves (though MEA is considered a neurotoxin).

However, once added to the product these chemicals readily react with any nitrites present to form the carcinogenic nitrosamine NDELA (N-nitrosodiethanolamine). Nitrosamines are among the major carcinogens in cigarettes and are also found in cured meats. In the 1970s nitrosamine contamination of bacon and other cured meats became a worldwide public health issue. As a result, the nitrosamine content of cured meats has dropped drastically in recent years. The nitrosamine content of toiletries, however, is alarmingly high. In a single shampoo you could absorb 50 to 100 micrograms of nitrosamine through your skin. A typical portion of bacon would only supply 1 microgram of nitrosamine.

Nitrites can get into personal care products in several ways. They can be added as anti-corrosive agents and they can be present as contaminants in raw materials. They can be the result of the presence of formaldehyde-releasing or formaldehyde-containing chemicals such as 2-bromo-2-nitropropane-1,3-diol (also known as BNPD or Bronopol) and Padimate O (octyl dimethyl PABA), DMDM hydantoin, diazolidinyl urea, imidazolindinyl urea and quaternium 15.

Some of the most commonly used ingredients that may contain DEA include Cocamide DEA, Cocamide MEA, DEA-cetyl phosphate, DEA olet-3 phosphate, Lauramide DEA, Linoleamide MEA, Myristamide DEA, Oleamide DEA, Stearamide MEA, TEA-lauryl sulphate and triethanolamine. The hair detangler stearalkonium chloride can also contribute to NDELA formation.

The long shelf life of most toiletries also increases the risk of creating a carcinogenic chemical reaction. Stored for extended periods at elevated temperatures, nitrates will continue to form in a product, accelerated by the presence of certain other chemicals such as formaldehyde, paraformaldehyde, thiocyanate, nitrophenols and certain metal salts.

Inadequate and confusing labelling means that consumers may never know which products are most likely to be contaminated. However, in a 1980 FDA report approximately 42 per cent of all cosmetics were found to be contaminated with NDELA – shampoos having the highest concentrations. In two 1991 reports 27 out of 29 products tested were found to be contaminated with NDELA.

While manufacturers plead that DEA and its relatives are safe in products that are designed for brief or discontinuous use, or in those that wash off, there is evidence from both human and animal studies that DEA is quickly absorbed through the skin. Also, this argument does not explain why these chemicals crop up regularly in body lotions and facial moisturisers, which are not washed off.

As far back as 1978 the International Agency for Research on Cancer (IARC) concluded that "although no epidemiological data are available, nitrosodiethanolamine should be regarded for practical purposes as if it were carcinogenic to humans". The IARC continues to maintain this position on NDELA.

In 1994 in America the National Toxicology Program similarly concluded in its Seventh Annual Report on Carcinogens that: "There is sufficient evidence for the carcinogenicity of N-nitroso-diethanolamine in experimental animals". They noted that of over 44 different species in which NDELA compounds have been tested, all have been susceptible, and that humans were most unlikely to be the only exception to this trend.

The response of the cosmetic industry to the problem of nitro-samiane formation in their products has been to put even more chemicals in the products we use in an attempt to slow or inhibit the formation on NDELA. None has been proved adequate against all possible nitrosating agents to which shelved cosmetics are exposed.

Carcinogens can get into your toiletries in other ways as well. Many of today's toiletries are based on compounds called ethoxylated alcohols. Ethoxylates can be contaminated with the carcinogen 1,4-dioxane. Commonly-used ethoxylates include PEG, polyethylene, polyethylene glycol, polyoxyethylene, 'eth' compounds such as sodium laureth sulfate or oxynol in the ingredients. Polysorbate 60 and polysorbate 80 may also be contaminated. In one 1991 study of a range of products including shampoos, liquid soaps, sun creams, bath foams, moisturising lotions, aftershave balms, cleansing milks, baby lotions, facial creams and hair lotions, more than half the products contained dioxanes at levels potentially harmful to human health.

• *But don't they do studies on these products to prove they are safe?*

This is a very common and fair question. If you were to ask representatives of the cosmetic and detergent industries whether this was so, they would reassure you that of course their products have been tested for safety. What is more, they would reassure you, those chemicals that have been shown to be toxic to humans are used in such small amounts that they could not possibly cause harm.

Such bland reassurances ignore the fact that it is often not single ingredients, but the combination of several chemicals in one product, that pose the greatest risk. While some ingredients have been tested for safety on their own, very little data exist on combinations of chemi cals. Nor are such data likely to exist, since the current production and use of chemicals greatly outstrips our ability to test them.

Worldwide, there are about 70,000 chemicals currently in day-to-day use, with 1,000 new chemicals coming on the market each year. To test the commonest 1,000 of these in combinations of three would require at least 166 million different experiments (and this disregards the need to test chemicals at varying doses). Even if scientists worked round the clock it is estimated that such data would take 180 years to compile.

Today a number of scientists have grave reservations about the chemical soups that we use on our bodies and in our homes. These mixtures can often react in unexpected ways, producing toxins that are harmful to human health.

In addition, when manufacturers try to reassure us that potentially harmful chemicals are only used in small amounts in their products they are glossing over some relevant facts. Every day we are exposed to many different chemicals. For example, we do not just use shampoo once in our lives, we use it regularly, sometimes daily. We use it in hot showers and baths where the chemicals can become vaporised and absorbed in greater quantities into our bloodstreams.

We use them in combination with other products such as conditioner, hair gel or foam, hair spray, toothpaste, deodorant, washing-up liquid, air fresheners and furniture polish all of which contain 'minute' amounts of the same harmful chemicals. Add up all these 'minute' amounts of chemicals and the potential for exposure is disturbingly high.

Finally, and perhaps most importantly, such bland reassurances ignore the fact that there is no basic safety information on 43 per cent of all the chemicals in use today and that full safety information is only available on 7 per cent of these chemicals. Among the chemicals commonly used in consumer products only 25 per cent have full safety data. It is frankly impossible for manufacturers to be equivocal about the safety of their products, given this appalling lack of data.

The EPA estimates that getting the full set of safety data on the most commonly-used chemicals would cost the chemical industry less than $427 million (£284 million). This represents 0.2 per cent of the total annual sales of the top 100 chemical companies (and a fraction of what is spent advertising the products made from these chemicals each year). Yet industry refuses to spend this money to provide such important data.

- *I don't suffer from sensitive skin or rashes when I use toiletries and household cleaners. Surely I'm not likely to be harmed by the chemicals in these products?*

Your body can respond to toxic exposures in two ways: either locally or systemically. A skin reaction is what is known as a local effect. With a local effect you can see the reaction and make a more or less accurate connection between the chemical (the cause) and the condition (the effect). In many ways those who get a local reaction such as a rash are lucky. Their bodies have given them a clear and immediate warning about the product.

But our bodies can also react systemically and it is this that causes the greatest concern amongst toxicologists and clinical ecologists. Many toxic chemicals are quick to enter the body, but slow to leave. Over the years levels can slowly build up in the body, weakening or damaging major organs such as liver, kidneys, lungs and thyroid.

When the body reacts systemically it is much more difficult to establish cause and effect. For example, if a person gets cancer, neither patient nor doctor would automatically assume that accumulated toxins in the body were responsible. In fact, from the point of view of most conventional physicians, cancer is a disease shrouded in mystery. But really there are few mysteries when it comes to cancer. Medical research tells us quite clearly that around 30 per cent of cancers have a genetic basis. The rest is down to environment – a broad category that includes dietary deficiencies and/or excesses and exposure to substances that are toxic to the body.

• *Which chemicals are the most dangerous?*

The most dangerous chemical in the house is the one that is not labelled. Far too many of the products we buy fall into this category because of inadequate or non-existent labelling of ingredients.

As far as this book is concerned, each of the sections that follows contains a list of the most harmful chemicals in specific toiletries and household products. Refer to those pages to learn more.

Chapter 2

Getting Beyond the Hype

More than 5,000 chemicals are commonly used in toiletries and household products. Thanks to the powerful machinery behind the manufacture and sale of these products, you probably have a number of them in your home right now.

Take a look at your bathroom. Typically it will contain bars of soap, body wash, shampoo, conditioner, bubble bath, body foam, hair spray, hair gel, styling mousse, toothpaste, eye gel, facial cleanser, facial toner, moisturiser, shaving foam, aftershave, deodorant and body spray, not to mention bath and tile cleaner, anti-mildew spray, toilet cleaner, air freshener and limescale remover.

And in the kitchen, typically, there is likely to be dish detergent, laundry soap, stain remover, fabric conditioner, floor cleaner, sink freshener, carpet shampoo, bleach, microwave cleaner, glass cleaner, stain remover, upholstery cleaner, fabric freshener, oven cleaner, hob cleaner, drain cleaner, floor polish, silver polish, brass polish, more air freshener and antibacterial spray.

Realistically, you need less than half of this stuff, but you buy it anyway. You may even rationalise your over-reliance on all these products by reasoning that 'they' would not sell it if it did not work and 'they' would not make it if it were not safe. But you would be projecting on to the manufacturers more responsibility than most of them actually take.

Products are invented in order to be sold. It is as simple as that. Since advertising companies have become more and more successful at 'positioning' products, i.e. defining their unique space in an overcrowded market, manufacturers have been falling over themselves to come up with 'new' and 'improved' goodies to fill these profitable niches.

To fill the perceived niche, every product you buy boasts its own specific formula – though in truth there is a limited number of variations on the recipe for any one kind of product. It is the responsibility of the formulation chemist to come up with that formula. Formulation chemistry is a specific branch of chemistry for which there is very little formal training. Most formulation chemists are self-taught or have learned their trade by working with other experienced chemists. The job of such people is to mix chemical compounds that do not react with each other in order to get a mixture with the desired characteristics.

This is where the job differs from, for example, an old-fashioned soap maker. Genuine soap is made from a combination of lye and fatty acids. The soapmaker relies on the chemical reaction between these two substances to create the end product. Although lye in its pure state could never be put on the skin without causing terrible harm, the chemical reaction of lye and fatty acids creates a new substance that is a safe and effective cleaner.

Modern products are more like suspensions of several compatible chemicals. However, it has come to light that even though these chemicals may not react immediately with each other at the formulation stage, they may react over the longer term, under non-laboratory conditions, to form harmful substances such as carcinogens (see chapter 1).

The work of the formulation chemist is very much focused on the short-term. He or she cannot use ingredients found on any 'banned' list (chemicals that are known to be overtly poisonous or dangerous) but beyond that he or she is free to use whatever produces a profitable product with the desired characteristics.

The final product, for instance, will have to smell nice, feel nice in the hands or on the skin, or look good in the bottle and after being poured out of the bottle. It may need to be mild enough not to cause immediate irritation on the skin. It may need to make use of fashionable new additives such as herbal extracts, natural smelling fragrances or fruit acids. It will need to do the job it is supposed to do, i.e. clean clothes, remove grease, or condition hair. It will need a reasonable shelf life, be somewhat resistant to changes in ambient temperature and be biodegradable. Beyond that the specifications for toiletries and household products (as far as human safety is concerned) are fairly narrow.

Once the product is formulated it is ready for rudimentary testing and then sale. Much of the safety of modern household products is defined by the use of ingredients that are generally recognised as safe (GRAS). These are ingredients that may have been tested sometime in the past or for which there is no data to show either safety or risk. Very few, if any, products are tested for long-term safety.

The next step is to get the products on the shelves – and then, hopefully, quickly off them again and into the shopping trolley. That is the job of the advertising agency.

Taking Ad-vantage

Pity the poor advertising man and woman. On their shoulders rests the fate of many industries and hundreds of thousands of jobs. If all advertising were banned today, within a week there would be no radio or TV stations and no magazines. Paper and film manufacturing companies would go out of business, people who mix ink, design costumes and direct TV shows would all be on the dole. Life as we know it would cease because we actually depend on advertising as a source of information about new products and as the motivating force behind purchasing, using and enjoying these products.

Keeping this massive industry afloat while not giving too much information away to the consumer is behind every media campaign for every product you use in your home. The consumer's role in all this is to sit passively, listen without interruption and not ask too many awkward questions. In short the consumer is expected to fill the role of a child – and frankly we do a pretty good job of it. As long as we do not drop dead on the first use of a product we assume it is safe. If we suffer from headaches, upset stomachs, or skin rashes, we shrug it off as 'normal' or 'to be expected' in this day and age.

Consumers are on the whole very undemanding – which is why manufacturers can get away with putting hazardous waste in their products and advertising companies can get away with ads that make these products sound like they are an absolute necessity and a privilege to buy.

How did we get to this state of affairs? In the earliest days of advertising, the goods on show really were necessities. Advertisers appealed to the common sense of the consumer with straight-forward information, e.g. 'You will need a warm coat for the winter.' At that time manufacturers could barely keep up with consumer demand. But at some point the position became reversed and the ability of manufacturers to produce products greatly outstripped the demands and the needs of the consumer. So, in order to keep buyers buying, advertisers had to learn to manipulate the emotions and the unconscious mind of the consumer. From very early on, they did this by creating fear.

A hundred years later the purpose of advertising is still to create fear. In fact, the shelves and cupboards full of toiletries and household cleaners in our homes are there almost wholly because we are afraid.

When people are afraid they are unlikely to be obsessive about seeking the truth or getting to the bottom of things. In this undemanding state we can easily allow ourselves to believe that something is really going to 'kill 99 per cent of household germs' or 'give visible results in just 2 weeks'.

Buying lots of stuff is our defence against fear. If you do not believe that, consider how often the language of warfare and defence is used in advertising. This kind of advertising tactic goes back to the Cold

War and the 1940s when petrochemical products were first introduced onto the market.

Since then we have come to believe that germs 'threaten' our families; they 'lurk' under the toilet seat and in the rubbish bin. So we 'fight' germs; we 'kill them dead', we 'get tough' on stains; we 'banish' odours. The bottles under the sink become part of the army we use to fight the faceless enemy of germs that are threatening to take over our homes.

We use the same language when it comes to our bodies. We use gel to 'control' our hair; we 'protect' our skin and 'take action' against the signs of ageing; we use mouthwash that 'kills the bacteria that can cause plaque'; we use antiperspirants to 'banish' wetness.

Even if the advert is blindingly funny, the underlying message is always a threat – often to your self-esteem. You are not a sexy woman unless you buy this shampoo or body spray. A real man uses this brand of shaving foam. You are not taking proper care of your family, or you are putting them at risk, unless you buy this anti-bacterial soap. Be careful, if your whites are not white, people will know how lazy you really are – and they will judge you.

Positioning products means giving them a unique identity. It is a sad fact that most consumer products have a better defined 'self' than many of the people who buy them. Thus, buying such products becomes a way of buying into a readymade identity and supporting a fragile sense of self-esteem.

For example, you may buy expensive designer brands or those with exotic sounding names 'because you're worth it', or the less expensive brands because 'you know it makes sense' and you are a 'sensible' person (this in spite of the fact that there is little difference in the ingredients between the two – remember it is all positioning). Or you buy the household cleaner that promises to work quickly without rinsing or the latest 'wash and go' shampoo because you are a fun-loving sort who knows that 'life's too short' to spend it cleaning or washing hair.

The final simple truth about advertising is: the more inferior the product the bigger and the harder the sell. Often the more dangerous it is for you to use, the more advertisements will try to convince you that it is good for you. Thus toilet cleaners that contain highly corrosive chemicals and are a major source of childhood poisonings, invariably feature cheery talking toilets singing their praises.

Actually you do not have to be too demanding to spot hype. But you have to be awake and aware in order not to allow yourself to respond to it. In addition to information on what is in the products you use everyday, this book offers many ideas for making simple alternatives from safe, basic ingredients.

Most people's immediate reaction to such make-it-yourself alternatives is that these are somehow second best. This reaction, which relies on emotion rather than common sense, is not entirely surprising. Multi-million pound industries spend many of their millions convincing us that we need them and their products to survive. If you feel like a poor relation using simple ingredients to get the job done, it is down to the invisible force of the chemical industry putting silent, but very effective, pressure on your psyche. For your health's sake, get past the hype.

Meaningless Terms

Manufacturers and advertisers use a variety of meaningless terms to convince consumers to part with their cash. To make conscious choices about the products you use, you may find it helpful to know more about such terms. Below are some of the most common claims that are used to baffle and distract the consumer from the real issues of safety and effectiveness.

Natural

The advertising world has co-opted the word 'natural' as one of its main selling tools. Many people feel greatly reassured when they see this word printed on a bottle – as if printing it automatically makes it so. There is no legal definition of 'natural' for personal or household products. From a chemist's point of view a chemical can be 'natural' or even 'organic' if its molecule contains carbon (the basis of all life on earth). Never mind that it may also be poisonous. From a manufacturer's point of view 'natural' can mean anything that sells the product.

Natural toiletries need only contain one per cent truly natural ingredients to earn the title. A product can even claim to be natural if what it really contains are 'nature identical' ingredients (in other word synthetic copies of natural substances). Some natural products can contain very unnatural substances and so abused is this word that we cannot legitimately point an accusing finger at the big corporations while seeking refuge in products from small manufacturers.

Go into any healthfood shop and take a closer look at their 'natural' toiletries and household products. By and large they are made from exactly the same ingredients as their more widely available conventional counterparts but are dressed up in reassuring language. A good example would be 'sodium lauryl sulphate (from coconuts)'. Don't you believe it – SLS is SLS. No matter what its origin it is still a harsh synthetic detergent. The same is true for all vegetable-based detergents, though these do have some environmental benefits over the petroleum-based variety.

Mild

Today mildness has replaced cleaning power as the major selling point of many types of detergent-based products. This trend has arisen out of the fact that detergents are some of the most common sources of adverse skin reactions. Because most detergents (especially household ones) are highly concentrated, none is ever truly mild. Instead most products are essentially a compromise. The formulators will still start with a strong detergent or surfactant such as sodium lauryl sulphate (SLS) and then combine it with milder detergents and other ingredients that modify the perceivable impact of the detergent on skin and hair. Ironically, some of the chemicals used to modify the harsher detergents are also common skin irritants.

Hypoallergenic

Many products now claim to be hypoallergenic. 'Hypo' means 'sub' or 'below' and the true meaning of the word 'hypoallergenic' is not 'allergen free' but 'lower in known allergens'. There are no regulations defining what an allergen is, though years of consumer complaint and dissatisfaction has defined many of the most common skin allergens in cosmetics, toiletries and household products. Equally, there are no official guidelines for producing hypoallergenic products. What is more, a product that is hypoallergenic may still contain substances that may be potentially carcinogenic or harmful to human health in some other way.

Fragrance-free

Fragrance allergy is the most common cause of contact dermatitis. Many individuals will invest in a fragrance-free product because they experience an allergic reaction to fragrance or because they are seeking to cut down on the amount of perfumes they use on a daily basis. The former group is the one most likely to be duped by the claim 'fragrance-free'.

Even fragrance-free products can have a recognisable scent. This is because while many fragrance-free products do not contain perfumes they can contain the raw ingredients of fragrance, often included to mask the odour of other chemicals. Such products can also include 'plant extracts', for example rose or tea tree oil or camomile, all of which have been found to be a considerable source of allergic reactions in sensitive individuals.

If you want to make sure a product has no fragrance you really have to read the label to make sure that it does not contain any plant extracts or other potential allergens. Going fragrance-free is also only half a solution for sensitive individuals. While most fragrance-free products are free of colour this is not always the case. Cosmetic colours are also a significant source of allergic reactions such as dermatitis.

Biodegradable

The biodegradable claim in soap powders and toiletries can be misleading on many levels, not least because it rarely represents an ethical decision and/or a change for the better in the development of such products. Almost any product can claim to be biodegradable as long as it eventually breaks down into constituent parts that are found in nature. The fact that these constituents could be toxic compounds such as mercury is immaterial.

The biodegradable claim is misleading for other reasons as well. Biodegradable products like food and leaves break down and decompose when exposed to air, moisture, bacteria and other elements. Other products like the plastic bottles in which your products are packaged are photodegradable, in other words they can disintegrate into smaller pieces if they are exposed to enough sunlight.

The landfills where most of our household rubbish ends up are designed, by law, to let in a minimum of sunlight, air and moisture. This, in theory, helps prevent the pollutants from entering the air and drinking water supplies. But it also greatly slows down the process of decomposition. Even organic materials such as paper and food may take decades to decompose in a landfill.

There is no government body monitoring the biodegradability of household goods such as detergents. The EU does not make any recommendations as to what ingredients synthetic detergents should or should not contain. In fact, detergent manufacturers seem to have been given a free hand to monitor themselves. While many manufacturers describe the basic cleaning agents on the label – many of which seem harmless enough and may well be biodegradable – they usually make no mention of colours, perfumes and preservatives, all of which could have a more damaging environmental impact once in the waste water system. Neither is there mention that certain detergents such as alkyl phenols can act as endocrine disrupters and once in the water supply become a threat to all animal life.

We buy biodegradable products because we assume that they are more ecologically friendly. As we look towards the future, however, we may need to be more demanding of our products. Not only should they be ecologically friendly, they should be biologically friendly as well. As biological beings, this should be our major priority.

Antibacterial

The antibacterial claim made on soaps and household detergents is another misleading one. A product can be called antibacterial if it contains ingredients that have been known to kill bugs in a laboratory. But we do not live in laboratories.

Ask yourself: How urgent is your need for special substances in your toiletries and household products to wash away dangerous bacteria? The germs you encounter in your home are unlikely to pose a threat to any member of your family – even babies and young children. Chances are you have already built up immunity to them.

Bacteria are programmed to survive. Mindlessly throwing anti-bacterial agents at them has been shown to make them stronger, not weaker. Hospitals, for instance, now find it impossible to kill all the germs in their environment – and they use some of the most power-ful antibacterial cleaners in the world. In hospitals food-borne diseases and antibiotic-resistant superbugs are now a major threat to health. Figures show that when you go into hospital you have as great as a one in 10 chance of coming out with a bacterial infection you did not have beforehand! Seriously, when the most powerful antibiotics in the most advanced hospitals in the world can no longer kill bacteria, what makes you think your dish detergent will?

pH Balanced

Contrary to what you have heard, skin and hair do not have an official pH though generally the skin is more acid than alkali. pH is measured on a scale of 0 (highly acid) to 14 (highly alkali) with 7 being considered neutral. 'Normal' skin pH ranges from 5–8.

Your skin produces keratin, fatty acids and other substances that work continuously to adjust its pH level. Almost anything you put on your skin, including water, will temporarily alter its pH. The pH of your skin can also change according to your environment and your state of health. Unless it is very harsh and applied continuously, the pH of a product will not alter the pH of the skin substantially or for long.

In reality there is no such thing as a pH-balanced product anyway. The product that left the factory with one pH may shift substantially during storage and shift again when applied to your hair or body and according to the pH of the water in which it is being used. Interestingly a product's ability to irritate the skin appears to be independent of its pH. While the pH 'balance' of any given product is unlikely to affect your body one way or another, the chemicals used to justify the claim 'pH balanced' can be irritating to your skin. Sodium hydroxide, or lye, is a good example of this.

CFC-Free
In the early 1980s chlorofluorocarbons or CFCs were banned from use as propellants in nearly all consumer aerosol products. Today hydrochlorofluorocarbons, or HCFCs, are sometimes used as a substitute for CFCs. While these are less damaging to the ozone layer, they still cause some ozone depletion.

Many commercial products on the supermarket shelves claim to be CFC-free or 'ozone friendly'. But you will need to look carefully at the label and think, not just of the ozone layer, which covers the upper atmosphere of the earth, but of ground level ozone as well. The upper layer of ozone protects us from the sun's harmful radiation. But when ozone develops at ground level, for instance as a result of the release of HCFCs, it forms smog that can cause serious breathing problems.

As another substitute for CFCs many aerosols and some pump sprays contain dangerous, volatile, organic compounds, or VOCs, such as butane, propane and isobutane all of which are serious neurotoxins. Read the label carefully if you want to keep them out of your home and your body.

Generally Recognised as Safe
GRAS – Generally Recognised as Safe – is an industry term, not a legal definition. The GRAS status of a product is usually determined by the

fact that is considered safe for use in foods. Many of the chemicals such as colours, flavourings (aromas) and preservatives, which are used in our toiletries and cleaning products, are also used in foods. However, the toxicity of a substance varies greatly according to the route of administration. No testing is required for other routes of entry into the body, and substances that are safe for ingestion, for example, may not be safe for inhalation.

The fragrance industry also uses GRAS status as an indication that the substance may be safely used to fragrance products. Yet, GRAS status requires no safety testing of a substance's effect on the skin, the respiratory system, or the nervous system. This is in spite of the fact that these pathways are the primary routes of exposure to fragranced products.

Chemicals in general use before 1 January 1958 were automatically given a GRAS status. But science has come a long way since then. We now know that many substances that were once considered safe are not safe. Yet they remain on the GRAS list and are widely used in many of the products we use daily. Current labelling laws are so inadequate that consumers will never know what they are inhaling or ingesting.

Not Tested on Animals

There is a genuine lack of information for consumers about the extent of animal testing in consumer products. Product labelling can often confuse rather than clarify the issue. You might buy a bubble bath which says that it is cruelty-free or not tested on animals and assume that you are doing your bit for animal rights. What that label probably means is that the *finished product* has not been tested on animals. The individual ingredients, however, may have been tested on animals either by the manufacturer or by an independent company hired to do the job or, more usually, may have been tested on animals many years ago before animal rights protesters got themselves organised into an effective group.

The thing to remember is that all of the available safety data on consumer products are based on animal data. And it is not just harmful chemicals that have been tested on animals. Somewhere in the world, at this moment, there are people testing garlic on animals or giving animals endless amounts of water to see what happens. There is little you can touch or see that has not involved cruel animal-testing in some way.

Do not become too comfortable either with cruelty-free natural products produced by small 'green' companies. Their hearts may be in the right place, but many of these products contain all the same old toxic chemicals that are in regular products. In addition, many of these products have built their 'cruelty-free' name on the backs of major manufacturers who have done the testing for them (and taken the flak for it).

Finally take note that the 'not tested on animals' claim does not refer to the widespread practice of post-marketing testing on the human animal.

Chapter 3

Wake up and Smell the Chemicals

Given the choice, would you prefer to live near a petrochemical plant or a perfume factory? Logically everyone knows that living next to a petrochemical plant would be dangerous. After all it would be spewing out hazardous waste and pollution known to cause serious health problems. What few of us know, however, is that modern perfumes are manufactured almost entirely from petrochemicals. Even more worrying, many of these chemicals are considered to be hazardous waste.

Virtually every aspect of our lives is touched by the fragrance industry. Watch a selection of TV commercials and you will soon become aware of the way in which the fragrance/aroma of a particular product is often given greater priority than performance. Fragrances are a multi-billion pound industry and the scary thing is that while evidence continues to accumulate on the harmful side effects of petrochemical fragrances, no one is regulating their use.

If perfume was once an art, it is now strictly a science, and a profitable one at that. Manufacturers long ago abandoned the use of genuinely natural ingredients. Instead synthetic fragrances are added to cosmetics and household products as well as foods – where they are called flavours or aromas. Flavours/aromas are also used heavily in the tobacco industry as additives in cigarettes to enhance flavour, especially in lower tar and nicotine brands, and to make the smell of second-hand smoke more acceptable (though no more safe) to non-smokers.

Several thousand different chemicals are used in fragrance manufacture; 95 per cent of these are derived from petroleum. Of the less than 20 per cent that have actually been tested for safety, most have been found to be toxic to humans. These include benzene derivatives, aldehydes and many other known toxins and sensitizers capable of causing cancer, birth defects, central nervous system disorders and allergic reactions.

Understanding our Sense of Smell

Odour and chemical detection is essential in the survival of most living things. In animals it is essential for finding food, for mating and for protection. In human babies the sense of smell is what helps them to find their mother's breast. But in adult humans its link to basic survival has decreased as our lives have become less dependent on being able to hunt and escape being hunted. Today we have scant regard for the importance of smell in our everyday lives. And yet our sense of smell is many thousands of times more acute than our sense of taste, and can still give us a great deal of information about our environments.

When you react badly to a smell you are reacting to a warning from your body that says you should not be breathing the chemicals in the air. However, if you are on a crowded train, or in an office or walking along the street you may not have a choice – you breathe or you die. So you ignore the warning. But the price of that denial can be chronic low-grade poisoning by volatile chemicals that can have wide-ranging effects on body and mind.

> The olfactory cells tire very easily. When exposed to an odour over an extended period of time, the perception of that odour becomes diminished. This is one reason why it is difficult to smell perfume on yourself, even when it is asphyxiating everyone else in the room. Each day we come into contact with a baffling variety of perfumes and scents, so much so that our senses eventually become overwhelmed and unable to process the information. When this happens they simply switch off. This dulling of our sense of smell may put us in a dangerous position, making it more much difficult for us to make the connection between fragrances in our environment and any health problems we may be experiencing.

The nose is a chemical receptor. When you detect an odour you are detecting the chemicals that make up that odour. When you breathe through your nose, volatile molecules find their way into your olfactory receptors. These receptors lead to a nerve pathway connected directly to the limbic system in the brain.

The limbic system is like a complex switchboard in your brain. Through this switchboard many bodily functions including emotion, instinct, voluntary movement, sleep, hormonal secretions and the interpretation of other senses apart from smell are regulated. There is nothing to stop the chemicals you inhale from reaching your brain. In fact they quickly and easily breach the 'blood-brain barrier' – the tightly-tiled lining of the brain's vascular system that allows energy-giving glucose to pass, but excludes most other substances.

As a general rule, product-labelling laws do not require manu-facturers to disclose what fragrances are used in their products and what they are composed of. This is considered either proprietary or a 'trade secret'. Yet, given that fragrances can quickly and easily gain access to the limbic system, and how wide-ranging their toxic effect can be, consumers should have an absolute right to know just what goes into their favourite perfumes and the fragrances used in toiletries and household products.

Health Effects

There is ample evidence that inhaled fragrances can cause immediate side effects. The fragrance portion of laundry products and cosmetics is also the number one cause of allergic and irritant skin reactions to those products. These are mostly local reactions caused by the product coming into contact with the skin. However, skin rashes can occur in sensitive individuals without the fragrance ever coming into direct contact with the skin.

Sometimes the adverse effects play out over the longer term. Chronic problems associated with fragrances include asthma,

What is in perfume? How bad can it be, you ask? Consider the ingredients for this 'timeless' light floral designer perfume:

Iso E Super, Lilial, Benzyl salicylate, Diethyl phthalate, Beta-ionone, Lyral, Alpha-terpineol, Piperonal, Galaxolide 50, Methyldihydrojasmonate, Linalyl acetate, CIS-3-hexenyl salicylate, Eugenol, Ethyl linalool, Cylcopentadecanolide, Linalool, Alpha-ionone, Benzyl acetate, Geranyl acetate, Octanol 7-hydroxy-3,7-dimethyl-, 2-Buten-1-ol, 2-ethyl-4-(2,2,3-trimethyl-3-cyclopenten- 1-yl)-, Benzeneethanol, 1,3-Benzodioxole, 5-(diethoxymethyl)-, 1-Cyclohexene-1-butanol 4-(diethoxymethyl)-.alpha.,.alpha.-dimethyl-, 2-Buten-1-one, 1-(2,6,6-trimethyl-1-cyclohexen-1-yl)-, Benzaldehyde, 4-hydroxy-3-methoxy-, Phenol, 2-methoxy-4-(1-propenyl)-, Phenol, 2-methoxy-4-(1-propenyl)-, 2-Octanol, 8,8-diethoxy-2,6-dimethyl-, 2-Propen-1-ol, 3-phenyl-, 6-Octen-3-ol, 3,7-dimethyl-, acetate, 6-Octen-3-ol, 3,7-dimethyl-, 7-Octen-4-one, 2,6- dimethyl-, Acetic acid, (cyclohexyloxy)-, 2-propenyl ester, 2,6-Octadien-1-ol, 3,7-dimethyl-, Phenol, 2,6-bis(1,1-dimethy-lethyl)- 4-methyl-, Benzaldehyde, 4-methoxy-, Benzenemethanol, 1,3,6-Octatriene, 3,7-dimethyl-, Benzoic acid, 2-hydroxy-, ethyl ester.

Romantic, isn't it? In the US an environmental group has petitioned the FDA to have this perfume declared misbranded. This particular perfume was chosen as an example because it is representative of the typical ingredients in all perfumes (its ingredients are also common to other fragranced products such as laundry detergents and toiletries). Of the 41 ingredients 33 (80 per cent) have no relevant safety data. Of the rest, data suggest that they are toxic, irritating, damaging to the central nervous system and carcinogenic. The basis of the complaint to the FDA is that none of the perfume's ingredients has been adequately tested for safety. Accordingly the group would like to see a warning on this (and all the other perfumes on the market) that says: *Warning – the safety of this product has not been determined.*

headaches (including migraines), spaciness, inability to concentrate, mood changes, dizziness, nausea, short-term memory lapse, restlessness, agitation, depression, sleepiness, lethargy and sinus pain. Some observers feel that a violation of the limbic system by volatile chemicals also plays an important role in multiple chemical sensitivity (MCS).

Studies have shown that inhaling fragrance chemicals can cause circulatory changes in the brain. Changes in electrical activity in the brain can also occur with exposure. Given this it is not surprising that chemically sensitive individuals have a higher rate of psychiatric disorders than others.

In addition to being inhaled, fragrances can be absorbed through the skin. The greater the emollient quality of the product you are using (think skin creams, roll on deodorants, etc.) the greater the absorbency. While fragrances can be quick to saturate the blood, they are slow to clear from the body. When fragrance chemicals penetrate the skin they can cause discoloration of internal organs. They can also be toxic to the liver and kidneys. Still others accumulate in fatty tissue and leech slowly back into the system or are passed on to our children through breast milk.

> We tend to perceive those substances that have a pleasant odour as good and those having an unpleasant odour as harmful. But this is not always the case. Many toxic solvents have a sweetish odour that is not unpleasant, but they can still cause serious harm to health. In 1991 the EPA found that toluene, a toxic volatile organic compound with a sweetish smell, was as common in the auto parts store as it was at the fragrance counter of upmarket department stores. Toluene can cause cancer and central nervous system damage and is designated 'hazardous waste' by the EPA.

Who is at Risk?

Women and children are most at risk from synthetic fragrances. Children and infants breathe at a faster rate, they are smaller and their sense of smell is more acute. In children, the rate of exposure in relationship to body weight is greater. Children's skin is also thinner which means that these substances are more easily absorbed. In short, when a child and an adult are exposed to the same toxin, the exposure is greater for the child.

A quick accounting of all those perfumed products we use on and around children is sobering: creams, ointments, talc, nappy sacks, laundry soap, fabric softeners, disinfectants, bubble bath, shampoos, sunscreens, soaps, menthol rubs and toothpaste are just a few examples.

In the spring of 2000 the Hyperactive Children's Support Group in the UK published an interesting survey. In it they concluded that perfumes and coloured products in the home were associated with significant adverse effects on children's behaviour.

They found that: 45 per cent of the children had reactions to perfumes; 80 per cent had a problem with colourings in medicines; 53 per cent had problems with the preservatives in medicines; 58 per cent reacted to coloured toothpaste; 41 per cent reacted to coloured bubble bath.

Women are also at risk. Their bodies are generally smaller and have a higher ratio of body fat than men's. Many toxic chemicals are stored in fatty tissue. In addition, women generally use more toiletries and household cleaners than men and are therefore more likely to be exposed to more fragrances. These fragrances are both inhaled and absorbed through the skin. Perhaps it is not altogether surprising that women who work at home are 54 per cent more likely to contract cancer than those who do not.

> Although the UK government announced its intention to investigate the presence of nitro musks in commercial products in 1994, these chemicals still remain on sale. Once in the body, nitro musks, sometimes referred to as xylenes, musk tibetene, musk ambrette, musk moskene, musk ketone and musk xylene, as well as some non-nitro musks are stored in fat tissue where they are very slow to clear. In the meantime they are carcinogenic and potential endocrine disrupters. They were commonly used in laundry powders up until a few years ago and although this use has been pretty much phased out, nitro musks continue to be used in soaps and other household and personal care products throughout Europe and the rest of the world. In a 1997 survey in Germany, out of 72 human blood samples, 66 (or 91 per cent) contained significant amounts of nitro musks.

The fact that smell and inhalation have such a direct connection to the brain goes some way toward explaining the devastating effects of glue sniffing and cocaine snorting. If we take glue sniffing as an example, we all know that children who inhale glue and solvents can become listless and violent, have poor concentration, experience convulsions and even coma. The solvents in glue are the same as the same ones used in fragrances. So what, you may ask, is the difference between wearing perfume and sniffing glue? Certainly there is a question of the concentration. But beyond that the answer is simply that the fragrance industry has a better PR than the teenage junkie.

Breaking the Fragrance Habit

We are enamoured with fragrances. We use them on our bodies, on our pets, on our children, on our furniture, on our clothes, and on most of the surfaces in our homes. This love affair with fragrance has allowed advertisers to pull the wool over consumers' eyes by linking fragrance with a desired quality such as love, sexiness, freshness, innocence and a wild, independent spirit. This message is so persuasive that some individuals feel they cannot be attractive unless they are wearing a scent. Our love affair with scent has even motivated us to believe in products that promise some sort of life-changing 'aromatherapy for your home'.

It is an unhealthy obsession, but not an inevitable one. Each of us has the capability to reduce the number of synthetic fragrances we come into contact with on a daily basis. If you need a little extra motivation to cut down, consider the documented health effects of some of the most commonly found chemicals in fragrance products.

- *Acetone* is on the 'hazardous waste' lists of several government agencies. It is primarily a CNS depressant that can cause dryness of the mouth and throat; dizziness, nausea, incoordination, slurred speech, drowsiness, and, in severe exposures, coma. It is found in cologne, dishwashing liquid and detergent and nail polish remover.

- *Benzaldehyde* acts as a local anaesthetic and CNS depressant. It can cause irritation to the mouth, throat, eyes, skin, lungs, and GI tract, causing nausea and abdominal pain. It has also been shown to cause kidney damage. It is found in perfume, cologne, hairspray, laundry bleach, deodorants, detergent, vaseline lotion, shaving cream, shampoo, bar soap and dishwasher detergent.

• *Benzyl acetate* is an environmental pollutant and carcinogen linked to pancreatic cancer. Its vapours are irritating to eyes and respiratory passages and it can also be absorbed through the skin causing systemic effects. It is found in perfume, cologne, shampoo, fabric softener, stickup air freshener, dishwashing liquid and detergent, soap, hairspray, bleach, aftershave and deodorants.

• *Benzyl alcohol* is irritating to the upper respiratory tract. It can cause headache, nausea, vomiting, dizziness, a drop in blood pressure, CNS depression and, in severe cases, death due to respiratory failure. It is found in perfume, cologne, soap, shampoo, nail polish remover, air freshener, laundry bleach and detergent, vaseline lotion, deodorant and fabric softener.

• *Camphor* is a local irritant and CNS stimulant that is readily absorbed through body tissues. Inhalation can irritate eyes, nose and throat and cause dizziness, confusion, nausea, twitching muscles and convulsions. Camphor is found in perfume, shaving cream, nail polish, fabric softener, dishwasher detergent and stickup air freshener.

• *Ethanol* is on the Environmental Protection Agency (EPA) 'hazardous waste' list. It causes CNS disorders and is irritating to the eyes and upper respiratory tract even in low concentrations. Inhalation of its vapours has the same effects as ingestion. These include an initial stimulatory effect followed by drowsiness, impaired vision, loss of muscle co-ordination and stupor. It is found in perfume, hairspray, shampoo, fabric softener, dishwashing liquid and detergent, laundry detergent, shaving cream, soap, vaseline lotion, air fresheners, nail polish and remover, paint and varnish remover.

• *Ethyl acetate* is on the EPA 'hazardous waste' list. It is a narcotic that is irritating to the eyes and respiratory tract. It can cause headache and stupor. It has a defatting effect on skin and may cause drying and cracking. In extreme cases it may cause anaemia with leukocytosis and damage to liver and kidneys. It is found in aftershave, cologne, perfume, shampoo, nail polish and remover, fabric softener and dishwashing liquid.

- *Limonene* is a carcinogen as well as a skin and eye irritant and allergen. It is found in perfume, cologne, disinfectant spray, bar soap, shaving cream, deodorants, nail polish and remover, fabric softener, dishwashing liquid, air fresheners, aftershave, bleach, paint and varnish remover.

- *Linalool* is a narcotic and causes CNS disorders. It has been shown to cause sometimes fatal respiratory disturbances, poor muscular co-ordination, reduced spontaneous motor activity and depression. Animal tests have shown it may also affect the heart. It is found in perfume, cologne, bar soap, shampoo, hand lotion, nail polish remover, hairspray, laundry detergent, dishwashing liquid, vaseline lotion, air fresheners, bleach powder, fabric softener, shaving cream, aftershave and solid deodorant.

- *Methylene chloride* was banned by the FDA in 1988 but no enforcement is possible due to trade secret laws protecting chemical fragrance industry. It is on the 'hazardous waste' lists of several government agencies. It is a carcinogen and CNS disrupter, absorbed and stored in body fat. It metabolises to carbon monoxide, reducing the oxygen-carrying capacity of the blood. Other adverse effects include headache, giddiness, stupor, irritability, fatigue, tingling in the limbs. It is found in shampoo, cologne, paint and varnish remover.

- *a-Pinene* is a sensitiser; damaging to the immune system. It is found in bar and liquid soap, cologne, perfume, shaving cream, deodorant, dishwashing liquid and air freshener.

- *g-Terpinene* causes asthma and CNS disorders. It is found in cologne, perfume, soap, shaving cream, deodorant and air freshener.

- *a-Terpineol* is highly irritating to mucous membranes; aspiration into the lungs can produce pneumonitis or even fatal oedema. It can also cause nervous excitement, loss of muscular co-ordination, hypothermia, CNS and respiratory depression and headache. Scientific data warn against repeated or prolonged skin contact. A-terpineol is found in perfume, cologne, laundry detergent, bleach powder, laundry bleach, fabric softener, stickup air freshener, vaseline lotion, cologne, soap, hairspray, aftershave and roll-on deodorant.

Note: CNS (Central Nervous System) disorders include Multiple Sclerosis, Parkinson's Disease, Alzheimer's Disease and Sudden Infant Death Syndrome.

Note: Seven other chemicals: 1,8-cineole; b-citronellol; b-myrcene; nerol; ocimene; b-phenethyl alcohol; a-terpinolene are among the 20 most commonly found chemicals in fragrances. There are no safety data on any of them.

Chapter 4

How Your Body Reacts

Polluted People

For years scientists have studied the effect of toxins on the environment. They have measured how much pollution is spewed into the air and dumped into the waterways. We now know a great deal about what happens to the earth, its natural habitats and the creatures that populate them when we pollute.

Perhaps it is in keeping with our arrogance and our insistence that we are separate from Nature, that we have never adequately studied what happens to humans when their bodies become polluted. Today this is an emerging field of medical and scientific endeavour.

As we have begun the process of studying polluted people, one concept has become pivotal – the 'total body load'. Understanding our total body load of poisons and toxins is the key to predicting and understanding the illnesses they cause. It is also the key to aiding recovery.

Our bodies are miraculous. They are equipped with many biological processes that help break down and excrete toxins from our systems. They do this automatically minute by minute and day by day. Skin, for example, is the largest excretory organ in the body. It helps to rid the body of toxins through sweat. It also protects us from dehydration and from temperature variations. The immune system protects us from foreign bodies such as bacteria that can make us ill. Kidneys eliminate waste products; the liver detoxifies the blood; the respiratory system oxygenates the blood as well as filtering out and excreting toxins.

There are, however, limits to what the body can do unaided. In order to continue to protect you effectively your body requires intelligent, practical help from you.

Certain lifestyle factors can make it much harder for the body to provide 24-hour protection. Often these are present in groups rather than singly. For example, a person may be eating foods that are laced with pesticides or to which they are intolerant; exposing themselves daily to toxins in the home or office; allowing emotional and psychological stress to build up; not taking exercise; not sleeping regularly; not allowing themselves to recover fully from illness before starting back to work; taking too many painkillers; drinking too much; and smoking cigarettes. This person's body is overloaded and will eventually break down. What breaks down first will be the weakest point in the system.

Each of us has some inherent constitutional weakness. These individual constitutional weaknesses are one of the reasons why it can be difficult to track the effect of the toxins in toiletries and household cleaners. For some the first symptoms may show in their digestive system, for others it is the skin, or immune system. Others may get headaches or symptoms of depression. Many potential effects of toxic exposures are discussed at greater length later in this chapter. However, two aspects of chemical poisoning – allergies and hormone disruption – deserve a closer look.

The Impact of Allergies

It is thought that around 15 to 20 per cent of the population experience significant allergic reactions to consumer products. An allergy, which is a response to something the body recognises as foreign or toxic, will usually produce physical symptoms such as skin reactions, digestive disturbances, headache or breathing difficulties. Mental and emotional disturbances are also common.

But in addition to these symptoms, an allergy will usually produce a measurable biochemical response. In other words your blood will have high levels of antigens – chemicals that stimulate an immune response. When you have symptoms, but no biochemical markers, this is called intolerance. With regard to toxic exposures it can be very difficult to tell the difference between an allergic or intolerance reaction and a reaction to low-level poisoning. Each has a similar effect on the body and some allergists believe that the distinction between them is merely academic, since the harm produced in the body is often the same.

Intolerance or low-level poisoning commonly causes symptoms such as joint pain, arthritis-like changes, chronic fatigue and a host of emotional states involving an inability to concentrate, feeling 'spaced-out' and woolly-minded, irritable, anxious or depressed.

To complicate things further, some allergies appear to produce no symptoms at all. This is a third type of allergic reaction known as the 'masked' or 'hidden' allergy. The concept of a masked and/or hidden allergy is controversial in some circles. In tracking down the causes of an allergy your average GP would look for obvious local symptoms (red skin, teary eyes). But an experienced allergy specialist would know that, in a case of hidden allergy, symptoms that may once have been acute become milder over time. They may even become diminished with greater exposure to the toxin, but the toxic effect of the substance on the whole of the body continues.

The hidden allergy is also similar to the effect of chronic low-level poisoning, in that the body adapts to the toxin over a period of time.

A simple example of the way the body adapts to toxins can be found in caffeine consumption. Caffeine (found in coffee, tea, sodas and chocolate) is one of the most commonly consumed psychoactive drugs in the world. When you first have caffeine it may make you feel jittery and irritable. But ingest enough of it over a long period of time and your body will adjust to the drug and the outward physical symptoms will begin to disappear. The caffeine, however, will continue to wreak havoc on your system by altering your metabolism, contracting your blood vessels and muting the response of your adrenal glands to stimulus.

It is not clear how many of the chemicals in everyday products may result in hidden allergies. Often such allergies can remain undetected for years, only revealing themselves as toxins are removed from the diet or environment and the body is given a chance to repair itself. Equally they can be revealed by some unusually large stress on the system such as:

Infection – Allergies can reveal themselves after a bout of severe viral, bacterial or fungal infection.

Chemical exposure – Examples include heavy exposure to pesticides or other petrochemicals.

Stress – An increased stress load, whether emotional or physical, positive or negative, can play a role in allergies. For example, marriage and birth can be happy events, but they are also stressful.

Nutrition – Poor nutritional habits and severe dieting can contribute to the development of allergies as well as other illnesses.

Symptoms of MCS often emerge after these types of stresses as well and, again, there is debate about whether MCS is the result of hidden allergies or chronic low-level poisoning.

Left untreated over the longer term, allergies can produce symptoms in almost every organ of the body, often masquerading as other diseases. Thus allergy can affect your skin, eyes, ears, nose, throat, lungs, stomach, bladder, vagina, muscles, joints, and your entire nervous system, including your brain.

With chronic exposure to toxins, the immune system may become so overworked and confused that it can no longer fight off further attacks. Or it may begin to attack itself as in the case of auto-immune diseases. Or the body may begin to produce abnormal cells as in cancer. Because the hormone and immune systems are so closely linked the person may begin to experience hormonal imbalances that can lead to thyroid problems, menstrual irregularities, or glucose intolerance and diabetes.

Many people resist the idea that they might have an allergy or that their bodies are reacting to low-level poisoning in an allergic-like way. But think about the medications that you take to deal with all those vague disorders such as sinusitis and headaches. Anti-histamines counteract the effect of histamine – a chemical released in the body when there is an allergic reaction. Aspirin and paracetamol are anti-inflammatory drugs. No one is sure exactly how these drugs work to relieve pain. However, inflammation is another common sign of allergy and poisoning. Even without thinking in terms of allergy, you and your doctor may automatically reach for anti-allergy medicines to deal with the toxic effects of chemical exposures.

Hormone Disruption

Another aspect of chemical exposure that has not been widely publicised but can still affect your long-term health, is hormonal or endocrine disruption.

When thinking about hormones it is important not to be limited by the popular view that hormones are only to do with our sexual and reproductive lives. Hormones are chemicals produced by the body that control a wide range of biological activities from metabolism to fetal development.

They are, for example, involved in regulating our moods and their disruption has been implicated in some emotional disorders such as depression. The endocrine system is also intimately related to other important body systems such as the immune system. Disruption in one can lead to disruption in the other and this disruption can in turn affect other body systems such as the nervous system.

A surprising number of synthetic chemicals (even those that do not closely resemble the structure of natural hormones such as estrogens) have the ability to disrupt hormonal function in animals, including humans.

Endocrine disruption is not considered to be a disease in itself. However, profound endocrine disruption can lead to diseases and disorders such as diabetes, hypothyroidism, cancer and reproductive or developmental disorders.

There are three types of endocrine disrupters: blockers, mimics and triggers. *Blockers* prevent the hormones produced by your body from getting where they need to go. *Mimics* act like natural hormones – estrogen is a commonly mimicked hormone. *Triggers* will do the same thing as blockers but will also trigger an abnormal response in the body.

When manufacturers defend the use of endocrine disrupters in their products they usually do so on the basis that such chemicals are only present in low doses and unlikely to cause harm. This argument has many holes. Firstly, the idea of low dose is usually based on experiments where a laboratory animal has been given a single or short-term exposure to the chemical. Because the chemical is not given over the longer-term or in combination with other 'low-doses' of known endocrine disrupters it is impossible to get a 'real world' picture of potential side effects.

Secondly, the endocrine system is remarkably finely tuned. Only minute amounts of a particular hormone, natural or otherwise, are needed to produce a reaction in the body.

Thirdly, the addition of a small quantity of any chemical to the body may have effects because it is acting in addition to another (similar) chemical that is already there. For example an estrogen mimic could act in addition to the natural estrogen already present in the body. Many cancers such as breast cancer and endometrial cancer are caused by 'too much' estrogen in the body.

Finally, it does not take into account the concept of synergism – the possibility that the combination of chemicals produces a response greater than a simple additive effect. In other words, when synergism occurs, it changes the natural order of things and two plus two suddenly equals five (or more).

Synergism is an important concept not just for endocrine disrupters but for carcinogens as well. It is estimated that combinations of chemicals can be anywhere from 160 to 1,600 times more toxic than they would be as single chemicals. We learned this lesson only too profoundly recently when it was found that the 'harmless' chemicals injected into Gulf War soldiers synergistically combined to produce powerful nerve damaging and reproductive effects.

Additive and synergistic effects are not limited to mixtures of synthetic chemicals and natural hormones in the body. Mixtures of synthetic chemicals can have additive and synergistic effects with each other (for instance while sitting on the supermarket shelf or in your cupboard) before they enter the body.

The science of mixture is so complicated and difficult to research that manufacturers have chosen to ignore it altogether. Nevertheless, it is a reality that urgently needs to be addressed.

Some of the chemicals known to disrupt the endocrine system include pesticides such as DDT and lindane, polychlorinated biphenyls (PCBs) and dioxins, fungicides and herbicides and bisphenol A, used to line tins that contain food. They are also found in heavy metals such as cadmium and lead (often used in pigment dyes) and mercury and in alkyl pheonls (nonoxynol, nonylphenol ethoxylate) such as those found in detergents used in laundry powders, shampoos in the US and heavy-duty cleaning agents in the UK.

The phthalates used widely in plastics such as PVC and found commonly in hairsprays and some conditioning shampoos, hair conditioners, deodorants (and in their plastic packaging) and fragrances are also endocrine disrupters. So are alkanolamines (a family of surfactants which includes DEA, TEA and MEA; see chapter 1) and parabens, a group of chemical preservatives used in a wide range of cosmetics, toiletries and household products and even food.

Lowering the Total Toxic Load?

Everyone can benefit from lowering the total toxic load on their bodies. But certain people are more susceptible to the adverse effects of toxic chemicals than others. If any of the following conditions apply to you, you may find that lowering the total toxic load on you body may help improve your health.

- Constant tiredness, chronic fatigue syndrome
- Auto-immune diseases such as osteoarthritis, diabetes, multiple sclerosis
- A family history of cancers
- Respiratory problems such as asthma or chronic bronchitis
- Sinusitis or hayfever
- Frequent colds, or susceptibility to flu
- Watery, itchy eyes or eyes that produce excess mucous
- Skin problems such as dermatitis or eczema
- Regular headaches or migraines
- Mental symptoms such as depression
- Trying to conceive; pregnancy or breastfeeding
- Menstrual problems such as PMS, irregular periods and anovulatory cycles
- Parasitic infestations
- Overweight or underweight.

Most of us would not automatically look for chemical causes to these conditions. While no one is suggesting that chemical overload is the sole cause of such problems, it cannot be ruled out as a contributing factor. What is more, diseases that do not seem to respond to conventional treatment often do respond once the environment and the body is cleared of toxins.

Taking a look at some of these conditions in more depth may help clarify the role of chemical exposures in each one.

Chronic Fatigue Syndrome (CFS)
When a person has chronic fatigue several body systems are adversely affected. These include the immune system, the endocrine system, the haematological system (responsible for the formation of new red blood cells) and the nervous system. Because of this CFS can produce a whole range of other symptoms, including headaches, low-

grade fever, swollen lymph nodes, sore throat, depression, poor ability to concentrate and decreased mental acuity, muscle and joint aches and pains, allergies, digestive complaints, weight loss, and skin rashes.

Mental and emotional symptoms are also common. CFS sufferers often have trouble remembering specific names or places or doing complex mental work such as bookkeeping, administrative tasks or teaching.

But severe fatigue is the most prevalent symptom, occurring in almost all sufferers. So much so that even minor exertion, such as a short walk or light housework, can be debilitating. While many sufferers try to curtail their normal daily activities and rest more, increased rest does not appear to help.

Many CFS sufferers report extreme and prolonged emotional stress, anxiety and depression, and a history of poor nutritional habits predating the onset of the condition. Environmental pollutants and contaminants are also thought to play a significant role in weakening the body and allowing CFS to develop. Having run the body down it becomes much more susceptible to viruses (the most widely accepted trigger of CFS).

Perhaps not surprisingly, given their higher rate of chemical exposure, women predominate among persons affected with CFS; more than 70 per cent of sufferers are female.

Autoimmune Diseases

The evidence – both anecdotal and scientific – for the link between chemical overload and autoimmune diseases is rapidly accumulating. For example, there is now good evidence to show that the auto-immune disease systemic lupus erythematosus (SLE) can be made worse by fluoride. Lupus is a connective tissue disorder. Connective tissue is made up of around 30 per cent collagen. Fluoride disrupts the synthesis of collagen and leads to the breakdown of this substance in skin, muscle, tendons, ligaments, bone, lungs, kidneys, cartilage and elsewhere.

Similarly allergic reactions are thought to play a part in the development of arthritis. This may be why the process of a medically supervised fast, followed by a low-toxin regime, can sometimes produce substantial relief of arthritis symptoms. Studies where arthritis patients have been exposed to various pollutants such as natural gas, auto exhausts, perfume, hair spray, insecticides and tobacco smoke have shown a clear link between exposure and a worsening of arthritic symptoms.

And while we have always assumed that diabetes is linked solely with diet, this may not be the case. There is evidence that chemical exposures during pregnancy may raise a child's risk of developing juvenile diabetes. New evidence shows that pesticides may also contribute to its development. Such findings have opened the door to the possibility that other types of chemical exposure may also be contributing factors to both juvenile and late onset diabetes.

Cancer

Cancer rates are on the rise. In the UK the number of cases diagnosed each year has increased by a massive 50 per cent in the past twenty-five years. Researchers say it is not just because the population is getting older. Nor is it likely to be that we are better at detecting the disease. So what is the cause? Cancer is not a disease that just suddenly develops. Instead, it is generated in two steps: initiation and promotion. The initiation process is what triggers the development of cancer. It is the factor or factors that interact with cellular DNA to start the process of producing abnormal cells. *Initiators* can be carcinogens (cancer-causing substances), viruses, radiation, oxygen, free radicals and hormones, particularly estrogens.

While none of these toxins may represent a significant carcinogenic exposure on their own, added together in susceptible individuals they become cumulative, challenging the immune system and damaging cells until eventually cancer develops.

Many of the ingredients in consumer products can act as initiators either because they are directly carcinogenic or because they disrupt hormonal or other normal bodily functions. These include volatile organic compounds and solvents, surfactants, formaldehyde, nitrosamines, plastics and preservatives.

Several lifestyle factors can also become triggers working away silently for years to damage cells. For instance, a low-fibre diet means that waste products remain in the gut longer, giving the body ample time to absorb toxins. Toxins in the body can interrupt the process of cellular repair, causing cells to replicate too quickly or to mutate.

After initiation, the disease may lie dormant for many years, until something comes along that allows it to grow. *Promoters*, which cause the disease to develop, are those things that damage the body's defence systems, particularly the immune system. Promoters may also alter body tissues, making them more favourable to cancer growth. A diet that is high in fat can make toxic chemicals much more efficient promoters. This is because chemicals can accumulate in fatty tissue where they can leech back into the body causing continual damage and eventually the development of disease.

Because cancer can lie dormant for many years before it begins to grow and spread, it can be difficult to link it conclusively with toxic exposures. Nevertheless, when cancer clusters (large numbers of people in one area with the same type of cancer) appear, environmental factors such as toxic exposures are always the leading suspect. It is estimated that between 10 and 30 per cent of cancers are the result of chemical exposures.

Asthma

Asthmatic attacks may be triggered by a variety of stimuli, the nature of which varies from individual to individual. Upper respiratory infections, either viral or bacterial, often trigger an asthmatic attack. Changes in weather or temperature, exposure to moulds, animal danders, grass or tree pollens are all triggers for some asthmatic patients. So are the preservatives, aromas and colourings used in pre-packaged foods.

Chemical exposures are also culprits. Once again perfumes are major triggers of asthmatic attacks. Around 70 per cent of asthmatics report that fragrance can trigger an attack. Other chemical triggers include the formaldehyde used in cosmetics such as medicated shampoos, germicidal soaps, mouthwashes and toothpastes. The solvents commonly used in toiletries and household cleaners can also cause respiratory disorders.

Skin Disorders

Skin rashes and other disorders of the skin constitute a large portion of diseases seen in general practice. Commonly encountered problems range from simple hives to impetigo to more complicated conditions such as eczema and psoriasis. The skin is the largest organ in the body and performs many life-sustaining functions. It is a barrier protecting us from harmful organisms in our environment, but it is also a major excretory organ. Skin reactions are commonly local reactions. However, they can also be as a result of the body trying to rid itself of toxins circulating deeper within the system.

The fragrance portion of toiletries and cleaning products is responsible for a high percentage of skin reactions. However, other chemicals are also implicated. These include preservatives, detergents and antibacterial agents.

While the skin can act as a barrier, this barrier is easily breached especially by solvents and other harsh chemicals. These can dissolve the protective mantle (composed of fatty acids) of the skin – and allow chemicals to be more easily absorbed directly into the bloodstream. Many chemicals actually aid the absorption of other more harmful chemicals into the body (see page 51). Skin permeability is also increased by hot water.

Headaches

If you get regular headaches it is likely to be because your body is under some kind of stress. For instance, the most common type of headache is the tension headache. While we often dismiss such headaches as the result of emotional stress, they are also likely to be the result of biological stress – such as food allergies and exposure to pollutants and toxins. Migraines are more complex, but also respond remarkably well to the removal of allergens and toxins.

Fragrances, in particular, are a major trigger of migraines. This is because fragrances can breach the blood brain barrier, gaining direct access to the central nervous system. Many other chemicals such as the solvents and propellants, that are so common in toiletries and household products, also affect the central nervous and vascular systems and are implicated in chronic headaches.

Research shows that pregnant women who constantly use aerosol products such as air fresheners, deodorants, furniture polishes and hair sprays have significantly higher rates of headaches and post-natal depression than those who use these products less than once a week. Research into migraines shows that smells are a common trigger.

Estimating how little or how much of a chemical actually enters the body is a complicated task. It is made more so by the fact that certain ingredients, while considered inert or GRAS, may enhance the skin penetration of other more toxic chemicals.

These include (among many others) the following common toiletry and cleaning agent ingredients:

Ionic compounds such as ionic surfactants, sodium lauryl sulphate, sodium carboxylate, sodium hyaluronate and ascorbate

Solvents such as acetone, ethanol, glycols, limonene, polyethylene glycol, propylene glycol, xylene, acetamide and trichloroethanol

Fatty acid esters such as butyl acetate, diethyl succinate, ethyl acetate, some isopropyl, methyl and sorbitan compounds

Fatty acids such as capric acid, lactic acid, linoleic acid, linolenic acid, oleaic acid and plamitic acid

Complexing agents such as lipsomes, naphthalenes, classical surfactants, non-ionic surfactants, nonoxynol, sodium lauryl sulphate and sodium oleate.

Depression

There is some evidence that environmental pollutants can cause or make worse feelings of depression, as well as other emotional problems. Many environmental practitioners have had good results in relieving depression by putting mild to moderately depressed patients on low-toxin regimes.

Depression also has links with immunity. Depressed immune function – often a result of toxic overload – can affect mood. Likewise a depressed mood can lead to depressed immune function, resulting in a vicious circle of poor mental and emotional health. Similarly, hormonal disruption can also be a factor in depression.

The chemicals involved include solvents, volatile organic compounds, and endocrine-disrupting chemicals. The fragrances used in perfumes, toiletries and household cleaners contain many of these and not surprisingly are often implicated in mood swings and emotional problems. Once you begin to limit your exposure to fragrances you may find that it is easier to recognise when a particular chemical ingredient or fragrance triggers a depressive episode.

Infertility and Reproductive Problems

Much of what we know about the influence of everyday toxins on fertility comes from research into those who work with these chemicals every day. For example, hairdressers and others who work in the beauty industry have much higher rates of infertility, miscarriage and babies with birth defects than women who are not exposed to these chemicals every day.

A recent Canadian study, published in the *Journal of the American Medical Association*, showed that women who work with organic solvents, e.g. artists, graphic designers, laboratory technicians, veterinary technicians, cleaners, factory workers, office workers, and chemists, have a greatly increased risk of both miscarriage and of giving birth to premature, low birthweight or damaged babies.

Organic solvents are not restricted to the workplace. They are also common in cleaning products, toiletries and paints. Pregnant women who live near petrochemical factories also experience more threatened abortions, toxaemia, anaemia, nausea and vomiting.

A man's fertility can also be affected by chemical exposures. Men who work in jobs which bring them into contact with pesticides and toxic chemicals are more likely to father children with birth defects. Such men often bring residues from their work home with them, increasing their partners' exposure to toxins as well.

Many studies have suggested that hormone-disrupting chemicals that find their way into the food chain are responsible for the global decline in sperm counts. Also some twenty different toxic chemicals with the potential to harm sperm quality and production have been found in random sperm samples. Contrast this with American evidence suggesting that men who eat organic foods have sperm counts that are twice as high as those who do not.

Finally, studies show that more than 350 man-made toxic chemicals are being passed on to babies in increasing amounts through pathways such as the placenta. Unlike adults, a developing baby is very sensitive to changes in the supply of nutrients and to the presence of poisons.

Perhaps not surprisingly, population studies tell us that the rate of birth defects has actually gone up in recent years. In the US one report showed that eighteen of the twenty most common birth defects were on the rise – some by as much as 1,700 per cent. Reports from the UK show a similar trend. This fact has been successfully obscured by an increase in medical terminations; as long as the baby is not born at term, the fact that it had an abnormality does not count in the 'official' records. Also in the West, problems such as undescended testicles have become so much more common that they are no longer recorded as malformations.

Yet, according to the National Network to Prevent Birth Defects in the US, a 50 per cent reduction in birth defects could be achieved if parents simply improved their diets and limited their exposure to toxic substances. The UK preconceptual care group Foresight has published figures showing that a pre-pregnancy health programme that includes dietary correction and reduction of toxins leads to an overall birth defect rate of less than 1 per cent – compared to 5 to 7 per cent which is the national average.

Parasites

Parasites latch on to those whose immune systems are functioning poorly. They can 'hibernate' inside the body as encysted colonies (resting and protected by a tough membrane that can resist most conventional medicines and pesticides). Generally, individuals whose immune systems are already compromised – those who undergo chemotherapy for cancer, those taking immune-suppressing drugs and those with AIDS – are especially vulnerable to parasitic infections (in fact, in these individuals parasites can be deadly). But because we are all daily bombarded with substances that over-stimulate and exhaust our immune systems we are all potentially vulnerable.

According to American Naturopath Dr Hulda Clark, solvents, so widely used in our toiletries and household cleaners, weaken the body and dissolve the tough membrane surrounding encysted parasites. This allows parasites to migrate to many sites in the body including the liver, breast, kidney and intestines. Once in these organs they can cause damage and may even act as promoters for cancer.

Many people recoil at the thought of parasites in the body, or believe that they are only a problem in the developing world. However, reviews of the literature show that parasites are no respectors of class or wealth. A random survey of stool specimens in the US, for instance, revealed that 20 per cent were infested with harmful parasites.

World-wide research into parasites estimates that at any given time 50 per cent of the world's population are infested with different classes of parasites. The eminent US physician Dr Leo Galland has gone on record with his belief that "every patient with disorders of immune function, including multiple allergies and patients with unexplained fatigue or with chronic bowel symptoms should be evaluated for the presence of intestinal parasites".

Eating Disorders

Both obesity and slimming disorders have been linked to toxins in the body. Continual ingestion of toxins can, for instance, produce nutritional deficiencies, particularly in zinc. Several studies show that zinc deficiency is a contributing factor to slimming disorders such as anorexia. Hormone disruption has also been linked to slimming disorders.

One German review has concluded that altered hormonal patterns are often diagnosed in young women before the onset of anorexia. Other researchers have found that one reason why bulimics report feeling uncomfortably full after eating may be irregularities in the hormonal process that regulates fluid volume in the body. Where this hormonal disruption comes from, and whether it begins at a young age with exposure to everyday toxins, may be useful avenues of research in coming years.

Similarly, some experts believe that obesity can be a response to accumulated toxins in the body. This may be due to the way toxins can alter thyroid function, making it more sluggish and unable to respond to the energy intake of the body.

However, it may also be because some of the most harmful toxins are absorbed by the body and stored in fat. As a rule the body does not like to let the concentration of potentially harmful substances rise too high and it will excrete them if it can. However, if for some reason the body cannot excrete them, or cannot excrete them quickly enough, the body's other option is to 'dilute' the toxins by storing them in fat cells. Under these circumstances, fat becomes the body's way of controlling and keeping high levels of toxins from circulating in the body.

During weight loss, as we lose fat, we release more toxins into the body. This may be why some dieters feel so headachy and unwell when they begin a slimming regime. To protect your body from the release of toxins while slimming you need a greater intake of water, fibre and the antioxidant nutrients such as vitamins C and E, beta-carotene, selenium and zinc.

Another way in which the body attempts to dilute toxins is through the accumulation of water. Individuals who feel that their extra weight is the result of water retention may also benefit from a sensible detoxification programme.

Detoxification
How can you help protect yourself and your family from everyday toxins? First you can become more discriminating and demanding about the products you use. Do not buy five cleaning products when one will do and be choosy about that one product.

In the chapters that follow you will see information on toxins in everyday products. This may stimulate you to use less hazardous alternatives and consider new measures to support your own body in the process of detoxification. You may decide to make some changes in lifestyle and diet. You may cut down on the number of products you use each day. You may also wish to think about ways of giving your home a thorough detox as well.

When you first stop using toxic products you may not notice much difference for a while, though some people may notice an immediate improvement. A good example of this is people who have been using baby oil to relieve dry skin. Often when they switch to a vegetable oil they notice that their skin becomes immediately less dry.

Other reactions are more subtle. It may be a month or so before you realise that you do not have a headache every afternoon at 4 pm, or that those persistent spots on your back have cleared up or that you have not had such severe asthmatic attacks recently. Most changes will be for the better. Within six months or so you may notice how sharply you react to the smell of disinfectant or perfume or other environmental toxins such as cigarette smoke. Similarly you may find that you are able to notice nice smells such as cut flowers more easily. You may realise with a flash how dulled your senses had become over the years.

But remember what you have read about hidden allergies. Often the withdrawal of one toxin or allergen will reveal a greater sensitivity to another. Take away the chemicals and you will suddenly become aware that certain foods or certain smells make you feel really ill. Heed the warning and cut down on these things or cut them out completely.

During the process of detoxification you need to follow your body's lead. For example, you may find you are more tired than usual. Sleep is when your body does essential maintenance work, so go to bed earlier for a while. Other ways to help your body include the following:

- *Change your diet.* Little changes can make a big difference. Switch to organic foods when you can and make sure your diet includes lots of fresh fruits and vegetables. The water-soluble fibre in these foods can help eliminate accumulated toxins. A diet that is low in saturated fat will also keep toxins from accumulating in the body. Try to favour oily fish over red meat since the essential fatty acids in mackerel, trout, salmon, herring and sardines can also help fight the effects of chemical exposures in the body.

- *Exercise.* Choose two or three activities that you really enjoy and work these into your schedule each week. Aerobic exercise or anything that makes you sweat is particularly useful for boosting metabolism and aiding the release of toxins through the skin.

- *Investigate stress relief.* The state of your body often reflects the state of your mind. Too much stress depletes the body of nutrients needed to fight the effects of pollution. Any activity that absorbs you completely and allows you to switch off can be useful in lowering stress levels. However, activities that involve deep breathing such as yoga or meditation can be very relaxing as well as promoting the release of toxins via the lungs.

- *Have a regular massage.* Consider booking an appointment for a regular massage. It is relaxing and can improve the movement of blood and lymph through the body, encouraging the removal of toxins, cell regeneration and better all-round circulation.

- *Try hydrotherapy.* Water in all its states helps purify the body. Try a relaxing bath with cider vinegar, epsom or sea salts added to stimulate the skin to release toxins. Invest in a home whirlpool for your bath. Or if you belong to a health club with a steam sauna, make regular use of it.

- *Use plant power.* Common houseplants can help remove toxins from indoor air. The best ones include aloe vera, English ivy, spider plants, peace lilies, philodendrons, bamboo plants, chrysanthemums, and daisies. Each has been shown to removes harmful formaldehyde, trichlorethelyne and VOCs such as benzene, toulene and xylene from the air.

PART 2

TOXIC TOILETRIES

Chapter 5

Bath Soaps and Body Washes

Soap or Detergent?

Soaps have been used for thousands of years as part of daily life and in religious ceremonies. However, today most of us do not use soap at all for anything.

Genuine soap is a simple substance made in a one-step process. It creates little waste in its manufacture and little waste in its use. To make soap all one requires is two ingredients: fat or oil and a strong alkali solution. Once mixed, the fat combines with the alkali solution to form soap. Good quality soap will contain around 70 per cent oil – for example olive, almond or coconut.

These days the art of making soap, and the practice of using it, has largely died away, overtaken by the trend towards detergent use. Today many liquid detergents and detergent bars are deceptively labelled 'soap' when they contain absolutely no soap whatsoever. This may sound incredible, but next time you buy a bar of soap look at the ingredients. They are likely to be synthetic detergents such as sodium tallowate, sodium palm kernelate or sodium palmate. If you buy a liquid soap or shampoo it may also contain ingredients like sodium lauryl sulphate or cocamide betaine or polysorbate 20.

The difference between soap and detergent is rather like the difference between cotton and nylon. Soap and cotton are produced from natural products by a relatively small modification. However, detergents and nylon are produced entirely in a chemical factory. Detergents have a greater impact on the environment than soaps, both from the waste stream they generate during their manufacture and in their poorer biodegradability.

This alphabet soup of ingredients constitutes a kind of chemical overkill and since many of these detergents are the same as those used in heavy industry, you have to wonder, just how dirty do manufacturers think you get each day?

A Brief History of Detergents

Detergents are part of a larger group of chemicals called surfactants (short for 'surface active agents'). Surfactants work by changing the properties of water. For instance, they can reduce the surface tension of the water, making it 'wetter' and better able to interact with other cleaning agents in the mixture. Detergents have similar properties to surfactants and may, in addition, add foaming ability. Both detergents and surfactants can be synthesised from either plant, animal or petroleum material. There is no difference between the detergents that are in your household cleaning products (see part 3) and those that you use in your bath. It is simply a matter of concentration.

The detergents that form a major part of most bath products and household cleaners were originally developed for industrial use in hard-water areas where they were thought to clean more efficiently. Since then research has shown that soap and detergent perform equally effectively in most types of waters although hard water appears to increase the potential of both types of cleaners to irritate the skin.

Detergent manufacturers also boast that, unlike soap, their products do not produce precipitate, i.e. the scummy substance that floats on the water or sticks to the side of the bath. But again this is not strictly true. All washing products produce some degree of precipitate – if not, what exactly have you been washing off the side of the bath all these years?

The problem can come during the rinse. In hard-water areas, both types of cleaners can be difficult to wash off; but old-fashioned soaps even more so. However, genuine castile soap, made with a high percentage of coconut oil, appears to rinse equally well in both types of water.

Having said this, even among detergents there is a great variation in both effectiveness and ecological impact. Those based on plant materials are somewhat kinder to the body and environment than those based on petroleum. And while for some industrial applications detergent is an appropriate choice, it is totally unnecessary, and potentially toxic, when used to clean the body.

Bath Bars

That good old bar in the bath or shower is the mainstay of most people's personal care regime. While they may all look the same, manufacturers claim significant differences with regard to the degrees of effectiveness and mildness between their products. This is one area where there seems to be some basis for such claims. For example, glycerine-based soaps are among the mildest on the market while deodorant and antibacterial soaps are among the harshest and most irritating to skin.

Somewhere in between lies the average bath bar which typically contains:

> sodium tallowate, sodium palmate, aqua, sodium cocoate, sodium palm kernelate, glycerine, coconut acid, parfum, pentasodium pentetate, sodium chloride, CI 77891, CI 74260, CI 47005, CI14700.

This bar contains several different synthetic detergents including *sodium tallowate* – made from animal fats – and *sodium palmate*, *sodium palm kernelate* and *sodium cocoate* which, although from vegetable origins, are purely synthetic chemical products that have none of their original vegetable characteristics. In addition it has:

- *Parfum* or fragrance of petrochemical origin. The fragrance portion of such products is a significant cause of skin irritation.

- *Colours*. In this case the white pigment *CI 77891* is titanium dioxide, a known skin irritant and potential carcinogen. *CI 14700* is known as FD&C Red 1. It is made with carcinogenic naphthalene and sometimes, depending on the formulation, aluminium. *CI 47005* is known as quinoline yellow or D&C yellow 10; it can also contain aluminium. *CI 74260* is known as pigment green for which there is no safety information.

- *Glycerine,* known technically as glycerol, is a form of alcohol used as a solvent and humectant (water magnet). It has a low toxicity but over time (like most humectants, including PEGs) it can paradoxically cause dry skin. Glycerine is often added to products like soap and glue to keep them from drying out.

- *Pentasodium pentetate* is an inorganic salt used as a water softener and preservative in products where metals are used. It can be irritating to the skin.

- *Stearic acid,* derived from both animal and vegetable sources, and *coconut acid* are fatty acids added as skin softeners.

- *Sodium chloride* is simple table salt, a water softener added to help the product rinse better in hard water.

In addition, many bath and hand soaps contain antibacterial agents. A US survey in 2000 found that 75 per cent of liquid soaps and one-third of bar soaps contained antibacterials. The picture in the UK is much the same.

Antibacterial ingredients are hard on the skin. But more importantly, they are linked to the emergence of antibiotic-resistant bacteria – a major health threat throughout the world.

Bubble Baths and Bath Foams
Of all the available bath products, bubble baths, which are highly fragranced, have the greatest potential to cause skin irritation, allergic skin reactions and headaches.

However, a bigger problem with using bubble baths is that they can irritate more than just your skin. Regular bubble bath use is associated with a high rate of urogenital infections. The harsh detergents in these products can strip away protective oils from sensitive areas of skin as well as stripping away the mucous that lines the genito-urinary tract. Removing this natural protection allows bacteria to take hold. Children are particularly vulnerable and bubble baths are a major cause of urogenital infections in babies.

Soaking in any bath product will prolong its contact with your skin. Hot water also increases your skin's permeability and helps vaporise some of the chemicals in the product so that they are more easily inhaled.

It is particularly important that you acquire the habit of reading labels. Otherwise you may end up soaking in a tub full of carcinogens. If your bubble bath has cocamide DEA (or similar compounds ending with DEA, TEA or MEA) along with formaldehyde-forming substances such as 2-bromo-2-nitropropane-1,3 diol (bronopol or BNP), DMDM hydantoin, diazolidinyl urea, imidazolindinyl urea and quaternium 15, you should not be using it since it is likely to contain cancer-causing nitrosamines.

A typical bubble bath product contains the following ingredients:

> aqua, sodium laureth sulphate, sodium chloride, parfum, cocamidopropyl betaine, cocamide DEA, citric acid, tetrasodium EDTA, benzophenone-3, methylchloroisothiazolinone, methylisothiazolinone, CI 42090, CI 19140.

This herbal foam bath, which encourages long soaks in order to soothe muscle strain, contains several chemicals that you might not want on or around your skin.

- *Cocamide DEA* is a strong detergent, foam stabiliser and thickener. It can irritate the skin. It also belongs to a family of fatty acids called *alkanolamines* that are considered hormone-disrupting chemicals.

- *Sodium laureth sulphate* (SLES) is a milder cleaning agent than its cousin sodium lauryl sulphate (SLS). However, SLES can be contaminated with the carcinogen 1,4-dioxane. There is no way of telling which products are contaminated and which ones are not.

- *Cocamidopropyl betaine* is a detergent. It is a strong allergen and skin irritant.

- *Methylchloroisothiazolinone* and *methylisothiazolinone* are preservatives and common skin irritants. It has recently been discovered that these two chemicals, which are the main ingredients in another preservative, kathon, have the potential to cause skin cancer.

- *Tetrasodium EDTA* or tetrasodium ethylenediamine tetra acetic acid, is a preservative used in soaps and other toiletries. It helps to isolate impurities such as metals that cause the mixture to degrade. It can be irritating to the skin and mucous membranes.

- *Benzophenone-3* absorbs ultraviolet light and is used as a sunscreen. Benzophenones are fixatives – chemicals that slow the rate of evaporation of the perfume component of a product. They are often found in soaps and detergents, which is why these products seem to be so much more strongly scented than others. Walk down the detergent aisle of your supermarket and you will smell how effective they can be. Prolonged skin contact with benzophenones may cause allergic reactions and lead to photosensitivity of the skin.

- *Colours.* The product under scrutiny contains the synthetic colours *CI 42090*, or FD&C Blue 1. It is a carcinogenic coal tar dye. It also contains another coal tar dye *CI 19140* otherwise known as tartrazine, or FD&C yellow 5, which can contain impurities such as carcinogenic heavy metals.

- *Sodium chloride*, or table salt, thickens the mixture and is also a water softener added to help the product rinse better in hard water.

In America bubble baths are now obliged to carry a health warning advising users to follow directions carefully and stating that prolonged use can cause skin irritation and raise the risk of urinary tract infections.

ℛ

Body Washes and Shower Gels

These come in gels and foams but essentially they are the same product as a bubble bath. A typical body wash contains:

> aqua, sodium laureth sulphate, socamidopropyl betaine, lauryl polyglucose, parfum, sodium chloride, lactic acid, tetrasodium EDTA, polyquaternium-7, benzophenone-4, citrus grandis, carica papaya, 2-bromo-2-nitropropane-1,3-diol, CI 16035, CI 47005.

This is one of those popular fruity body washes that are so heavily touted as being 'nourishing' for your skin. However, it contains the preservative *2-bromo-2-nitropropane-1,3-diol* which is a formaldehyde-forming chemical (formaldehyde is a carcinogen) as well as the skin-irritating detergent *cocamidopropyl betaine*. It also contains synthetic *fragrance* and coal tar colour *CI 47005* D&C yellow 10, which can contain aluminium and *CI 16035*, or FD&C red 40, an azo dye that is made with naphthalene (a carcinogenic solvent) and may be contaminated with carcinogenic impurities such as heavy metals.

It contains the detergent *sodium laureth sulphate* (SLES) that can be contaminated with the carcinogen 1,4 dioxane and *benzophenone* and *tetrasodium EDTA* (see bubble baths). *Polyquaternium-7* is a quaternay ammonium compound used as a surfactant, conditioner, thickener and emollient. Quaternay compounds are found just as often in fabric softeners as they are in products that promise to soften the skin and hair. They can be irritating to the skin.

Fruit acids (alpha hydroxy acids or AHAs) such as those from citrus (in this case grapefruit or *citrus grandis*) and papaya (or *Carica papaya*) are popular among those who like to think of their toiletries as 'natural'. However, they have the potential to irritate and harm your skin. *Lactic acid* is also an AHA. Some skin products contain beta hydroxy acids or BHAs such as salicylic acid and glycolic acid which are considered somewhat less irritating to the skin than AHAs.

The word 'acid' should give you a clue as to how these particular chemicals work. In their more concentrated form AHAs are used by dermatologists to burn off the top layers of their patients' skin (usually as a treatment for skin diseases and wrinkles). Patients like to call such procedures 'facial peels'; dermatologists call them 'chemical burns'. While not in skin products at anywhere near this concentration, the effect of AHAs is still to burn the skin slightly. The 'glow' you get from products containing AHAs is actually a mild irritation caused by the chemical burn. Some people react more violently to this burning than others and AHAs are reported to be a major cause of adverse reactions to soaps, shampoos and moisturisers.

Alternatives
Most bath products are frankly unnecessary. Anything that produces a lot of foam has been made to appeal to your sensuous nature, rather than to your desire to be clean. The best alternative is to stick to bath bars, avoid bubble baths altogether and limit your use of bath foams and shower gels.

If you are looking for the mildest way to clean your skin, you should follow these steps when you shop:

• *Always opt for vegetable-* and glycerine-based soaps over harsher petrochemical-based varieties.

• *Buy real soap* made from at least 70 per cent vegetable oil. Many health food shops stock such soaps or you can order them from specialist suppliers.

• *Choose a liquid castile soap* instead of a body wash. Liquid castile soaps (such as those made by Dr Bronner) foam beautifully and are made from enriching oils such as coconut, hemp and olive. They are usually (but check the label) fragranced with essential oils and even come unscented so you can add your own fragrance.

- *If you must have bubble baths* have them less often and make sure you take them in a well-ventilated room to avoid inhaling too many chemicals.

For the more ambitious, making your own bath products means you can have exactly the right scent to suit your mood on any given day. It also ensures that you are not soaking in or inhaling a bathtub full of potentially harmful chemicals.

- *Use essential oils* to fragrance your bath. To help them disperse in the water mix 4–10 drops of essential oil in 15mls of milk (semi-skimmed or whole). The fat in the milk will distribute the oils evenly around the bath. Alternatively, mix in a carrier oil such as almond or grapeseed.

- *Have a fragrant mineral bath.* Add a generous handful of Epsom salts and a few drops of your favourite essential oil (mixed in a carrier as above). Epsom salts can be purchased at any chemist in large 3kg bags for a minute fraction of the cost

---

For a bit of bath therapy choose oils that match your skin type. For example:

Greasy: Lavender, orange, lemon, clary sage, neroli, cypress, ylang-ylang, bergamot

Normal: Palma rosa, geranium, lavender, Roman camomile, jasmine, neroli, ylang-ylang, frankincense, sandalwood, patchouli

Sensitive: Geranium, lavender, German camomile

Dry or damaged: Geranium, lavender, German camomile, Roman camomile, clary sage, naiouli, thyme linalol, myrrh or a mixture of eucalyptus and lemon or peppermint.

---

of name brands that are essentially the same thing. For a bit of a fizz in the bath add a handful of bicarbonate of soda as well. Epsom salts are a good way of encouraging the skin to release accumulated toxins, so in addition to being pleasant and safe to soak in, they are therapeutic as well.

- *Make your own bathbomb.* Commercially made bathbombs may contain dubious chemicals and colours. Instead if you want your bath to fizz nicely, mix 3 tblsp (45mg) bicarbonate of soda with 1½ tblsp (22mg) of citric acid in a bowl together with 8–10 drops of your favourite essential oil. Drizzle a scant teaspoon of water over the dry ingredients and mix well. This is enough for one large bathbomb. You can press it into a mould (an old film cartridge or an ice tray will do) and store in a plastic bag for later. Or you can use it right away by sprinkling the mixture in the bath just as you are getting in.

- *Have a herbal bath.* Brew up a strong infusion of your favourite herbal tea and mix this into your bath water. Good choices include peppermint, camomile, lavender and limeflower.

- *Make your own herbal wash bag* if you suffer from dry, irritated skin. Cut the foot off an old pair of tights, about 6 inches from the end. Fill the pouch with a handful of oatmeal, some soothing herbs such as camomile or lavender and 2 tblsp of finely ground almonds. Tie a knot in the open end of the pouch. You can now use this in the bath or shower. One wash bag will last one day maximum – keep it in the fridge in a plastic bag if you intend to use it morning and night, but do not try to store it longer than this as it can accumulate bacteria. When wet it will produce a lovely creamy (soap-free) liquid that will clean and nourish your skin without drying it. This is great for adults but also for babies and children with dry skin.

Chapter 6

Dental Care

Toothpaste

Today there is no such thing as simple toothpaste. Instead, like so many other toiletries, it has become a magic potion. Believe what you see on TV and toothpaste has ways of making you more attractive to the opposite sex, 'protects' your family from the unseen but persistent threat of germs and disease and keeps your body under control so that it will not betray you at the wrong moment with bad breath.

Amazingly, in the UK anybody can make and sell toothpaste. As long as the product does not make any medicinal claims (such as those that reduce tooth sensitivity) it does not even need a product licence.

While there is a wide variety of toothpastes on the market, most are basically the same – there is no advantage of gel toothpaste, for example, over cream formulas. And while many toothpastes appear to make grand claims about what they can do for your teeth, look closely at the wording on the box next time you buy. It will not actually say that it will prevent plaque or tartar build-up. Instead it will claim that it 'fights it' or 'can help prevent' it. This is a convenient way of sounding like a medicinal claim (and thus encouraging the trust of the consumer) when it actually is not.

Tartar-control toothpastes cannot remove existing tartar. Likewise, those pastes that claim to prevent gum disease cannot treat existing gum disease. In both cases a trip to the dentist is necessary. Smoker's toothpaste, which claims to remove stains, can be highly abrasive. Excessive abrasion on the teeth can damage enamel and cause the gums to recede. Sensitive tooth formulas contain nerve-deadening chemicals such as *strontium chloride* or *potassium nitrate*. Do not expect these toothpastes to work immediately, like they do on TV – most people won't notice any improvement for 4–6 weeks. If your teeth are very sensitive you will need a trip to the dentist to rule out an underlying problem, such as a cracked tooth or gum disease.

Looking at the label on a typical toothpaste tube you might wonder why you would want to put this product in your mouth at all. In a market crowded with alternatives each brand has to make bigger claims about what it can do. And to back up these claims more and more chemicals are added. Should you be worried? Probably.

Consider the following typical toothpaste label:

aqua, sorbitol, hydrated silica, glycerine, tetrapotassium pyrophosphate, PEG-6, tetrasodium pyrophosphate, sodium lauryl sulphate, disodium pyrophosphate, aroma, carbomer, dipotassium phosphate, sodium phosphate, sodium fluoride, sodium saccharin, triclosan, xanthan gum, Cl 74260, Cl 77891 Contains pyrophosphate 5%, 0.32% sodium fluoride, 1450ppm fluoride.

In this example there are several worrying ingredients:

- *Sodium lauryl sulphate (SLS)* is a detergent and foaming agent. It is added mostly to help toothpaste feel nicer in your mouth. If you are prone to canker sores or mouth ulcers, consider SLS as a potential culprit because it can be highly irritating to the delicate skin around the mouth. SLS is certainly not necessary to clean your teeth.

- *Hydrated silica* is used mainly as an abrasive. While it is the powdered form of silica (found in cosmetics and scouring powders) that is considered carcinogenic when inhaled, this paste contains these same particles. The safety of silica through other routes of ingestion has not been adequately proven.

- *Triclosan* is an antibacterial agent found in toothpastes, face washes and bath products. It can cause allergic reactions and ulceration. It is easily absorbed into the body via the mouth, and has been associated with liver damage and eye irritation.

- *Pyrophosphates* are anti-plaque ingredients. There are several documented problems that come from adding pyrophosphates to toothpaste. First, pyrophosphate makes the mixture slightly alkaline. Once in the mouth, which is naturally slightly acidic, it can irritate the mucous membrane. Second, pyrophosphates taste terrible so more flavouring (aroma) needs to be added to mask the taste (see below). Third, pyrophosphates do not dissolve easily, so more detergent needs to be added to the mix to help dissolve it.

- *Carbomer* is a gelling agent. It is a synthetic polymer (a plastic-like material) used to thicken, stabilise and promote the shelf life of the product. It can be irritating to skin. *Xanthan gum* is also used as bulking agent, though it has no known toxicity.

- *Saccharin* is a carcinogenic artificial sweetener. Note the product also has *sorbitol*, an alcohol-derived sugar that tastes sweet but only breaks down in the gut. If swallowed, sorbitol can cause gastrointestinal cramps and bloating.

- *Aromas (flavours) and colours* are added to make the product look, taste and smell more appealing; they do not help clean your teeth. Aromas are generally petrochemical-derived. In some products additional chemicals are added to help the aroma be released faster in the mouth – they have nothing to do with cleaning either. Colours are there to make the product more appealing to the eye. In the example above, the colours added are *CI 77891*, titanium dioxide, a white pigment and powerful skin irritant shown to cause tumours in experimental animals, and *CI 74260*, a green pigment for which there is no safety information.

The F-word

More worrying than any of the above ingredients, however, is the fact that nearly all toothpastes on the market contain fluoride – a widely-used insecticide and acknowledged poison.

In the chemical tables fluoride falls somewhere between arsenic and lead in terms of its toxicity. When containers of fluoride arrive at the doors of toothpaste manufacturers they do so with a skull and crossbones on the front. That picture is a far cry from the happy (because their teeth are 'safe' thanks to fluoride) smiling faces we see in toothpaste advertisements.

Strangely there is very little conclusive evidence that fluoride in toothpaste or in water has directly led to better dental health. Much of what we know is based on assumption. Some investigators have been more forceful in their condemnation of fluoride, believing that most of what we know is based on governmental lies and deception.

Toothpastes contain variable amounts of fluoride – anywhere from 100ppmF (parts per million of fluoride) for a baby toothpaste to 1,500ppmF for an adult one. Looking at these figures, the average consumer might ask, what do the numbers actually mean? It is a good question, made harder to answer by the fact that in the UK labelling information is not consistent from product to product. Some companies list the amount of fluoride in ppms and some in percentages, some in both and some not at all.

The British Dental Association (BDA) recommends that all toothpastes should list their fluoride content in ppms because this is the easiest way to help consumers figure out the total amount of fluoride in any given product. In the US ppm is already a standard on toothpaste tubes (as is information on how to contact the local poison control unit, should the product be ingested).

Toothpastes that list fluoride content in percentages instead of ppm are very misleading because these figures do not refer to the actual amount of fluoride in the tube. Instead they refer to the amount of the chemical compound of which fluoride is a part. Sodium fluoride, for example, is a compound containing sodium and fluoride (with fluoride making up approximately 45 per cent of the mixture); *Sodium monofluorophosphate* is a compound of sodium, fluoride and phosphate (containing about 10 per cent fluoride).

Probably toxic

Several recent reports show that what is known as the 'probably toxic dose' (PTD) of fluoride, i.e. a single ingestion that would require medical intervention and hospitalisation, is actually much lower than previously thought, between 3.2 and 6.4mg of fluoride per kg of bodyweight. This of course is only a guideline and acute poisonings have been reported at doses of 0.1 to 0.8mg/kg of body weight.

Many widely-used brands contain sufficient fluoride to exceed the PTD for young children. For instance a 10kg child who ingests 50mg fluoride (roughly equivalent to one-third of a 100ml tube of 1,500ppm toothpaste or half a 100ml tube of 1,000ppm toothpaste) will have ingested a probably toxic dose and will require hospitalisation. Eating half a tube of 1,500ppm paste (and it happens more frequently than most parents would imagine) could kill the child.

How much fluoride?

To figure out the amount of fluoride in your toothpaste, look for the ppm. If your toothpaste tube only lists percentages, use the chart below to find the approximate ppm of your toothpaste:

ppm	Sodium Fluoride	Sodium Mono-fluorophosphate
1,500	0.32%	1.14%
1,000	0.22%	0.76%
500	0.11%	0.38%

To find out how many milligrams of fluoride are in the tube divide the ppm by 1,000 and then multiply that figure by the weight of the tube. Thus: 1,500ppm ÷ 1,000 = 1.5 x 100 = 150mg.

As a rough guide, a 100ml tube of 1,500ppmF toothpaste will have 150mg of fluoride in it. At 1,000ppmF there will be 100mg of fluoride and at 500ppmF there will be 50mg.

Too much fluoride in toothpaste for young children can also lead to fluorosis – a permanent staining and mottling of their teeth. Young children have a tendency to swallow toothpaste and with it the toxic fluoride. This may be because they are too young to control their swallowing or it may simply be that sweet flavours and pretty colours make toothpaste as appealing as candy to swallow.

For both these reasons it is now widely accepted that toothpaste with 1,500ppmF fluoride in it is unsuitable for children under the age of 8.

Why so Much fluoride?

Toothpaste is rinsed out of the mouth with water, and it is thought that teeth only retain a small proportion of the fluoride in toothpastes and mouth rinses.

Also, fluoride can be unstable. Depending on the formulation, the amount of fluoride in toothpaste can decline over time. For instance, when sodium fluoride is combined with aluminium- and/or calcium-containing abrasives the mixture will lose between 60–90 per cent of the added fluoride after just one week's storage at room temperature. These forms of abrasives are not often seen in big brands these days – but calcium-based abrasives are widely used in 'natural' alternatives, many of which also contain fluoride.

To overcome these problems manufacturers just add more fluoride.

Whitening Toothpastes

Recently there has been a trend towards using toothpaste that promises to deliver a whiter smile. Such toothpastes are popular among smokers and among those with low self-esteem. These toothpastes work in two ways, either by bleaching the teeth with chemicals such as hydrogen peroxide (which can be a slow process) or by using gritty particles to rub away stains. The abrasive action is harmful to your teeth. The jury is still out on the chemical bleaching, but certain problems have been observed.

At best these toothpastes may produce mild cleaning or bleaching action that only lasts as long as you keep using the product. At worst they can make your teeth painfully sensitive to heat and cold, may irritate your gums and permanently destroy tooth enamel. Concern has also been expressed that new chemical whiteners are still something of an unknown quantity. Peroxide, for instance, is highly reactive and may interact with other chemicals present in the paste to form new, harmful chemicals.

Mouthwash
The traditional purpose of a mouthwash is as an antiseptic gargle to help remove germs that can lead to bad breath. Today mouthwashes come in many more complex forms that purport to fight plaque, strengthen teeth, fight tooth decay and freshen breath in addition to killing germs. The result is a complex mixture of chemicals that, in the long run, may do more harm than good.

A typical mouthwash contains:

> water/aqua, alcohol, flavour/aroma, cocamidopropyl betaine, sodium saccharin, cetylpyridinium chloride, codium fluoride, PEG-60, hydrogentated castor oil, sodium bicarbonate, CI 42051, CI 75810. Contains sodium fluoride (225ppm fluoride).

No matter how careful you are, when you gargle with a mouthwash some of it will be ingested. Also, because mouthwash residue is left in the mouth, rather than rinsed out with water, you may be ingesting more than you realise.

> Breath sprays are concentrated forms of mouthwash. Typically they contain more alcohol than water as well as isobutane, glycerine and other sweeteners such as saccharin or sorbitol, flavourings and even colours. Because they are so high in alcohol and they are not rinsed out of the mouth, regular users are at increased risk of throat cancer. Using a spray will not stop you having bad breath.

Although mostly water, there are several problems with the current crop of mouthwashes. Firstly, most contain *alcohol*. Use of alcohol-containing mouthwashes is associated with an increased risk of throat and mouth cancers. This is because alcohol is drying, changes the pH of the mouth, and strips away the protective mucous membrane in the mouth and throat. Although in a low concentration, most mouthwashes also contain *fluoride*. Drinking alcohol- and fluoride-containing mouthwash is a major source of poisoning among young children.

The ingredients also include the following:

- *Cetylpyridinium chloride* is an antiseptic and preservative. It has never been adequately proven to kill germs but it can cause teeth staining, a burning sensation in the mouth and increased tartar formation.

- *Artificial sweeteners* such as saccharin, but also sorbitol, xylitol, manitol and cyclamates. These are not necessary for cleaning and if ingested can cause stomach upset and bloating. Saccharin is not used as a sweetener in the US because it has been shown to be carcinogenic at high doses.

- *Harsh detergents* such as *cocamidopropyl betaine*. These are also responsible for stripping the mucous membrane and altering the pH of the mouth and have been implicated in allergic reactions.

- *Colours*. In the example above the manufacturers used synthetic food colouring Patent Blue V (*CI 42051*) and a natural green known as chlorophyllinium cupreum (*CI 75810*).

- *Preservatives*. PEG compounds can be contaminated with the carcinogen 1.4-dioxane.

- Ingredients like *castor oil* and *sodium bicarbonate*. These are included to improve the feel of the mixture.

Typically, the example above did not specify which *flavouring (aroma)* it used. Flavourings, of course, are not effective breath fresheners. At best they can only mask odour for a short time. Mostly they are included to boost the perceived 'freshness' and effectiveness of the product. Strong flavourings such as *methyl salicylate* (oil of wintergreen, a common flavouring in mouthwash) may smell pleasant but they are highly toxic. Just 10ml is toxic to a child and 30ml is toxic to an adult.

With good oral hygiene a mouthwash is never necessary. A healthy mouth simply has no odour. If you are experiencing a problem with persistent bad breath it could be because of gum disease or some other underlying infection. This problem is most effectively addressed by a trip to the dentist.

The Low-down on Dental Floss

Although we think of flossing as a recent development in dental care, evidence gleaned from the skulls of prehistoric humans suggests that we have been flossing in one way or another for a very long time. In the 1800s floss was made of fine unwaxed silk thread. Today it is made of thin nylon string.

Some holistic dentists have expressed concern that modern dental floss can be contaminated with mercury-containing antiseptics. Certainly some flosses are impregnated with unnecessary flavourings derived from petrochemicals and some are even coloured. Often they are coated with waxes of petrochemical origin as well.

American Naturopath Dr Hulda Clark has gone so far as to recommend that flossing is best done with a 2 or 4 pound monofilament fishing line (doubled and twisted for strength)! She also suggests that the thin plastic used for most shopping bags is also serviceable. Tear off a thin strip and roll it slightly to make a quick, effective floss.

If you do not fancy that or do not have a fisherman in the family you might just try to be a little more aware of the types of dental floss you use. Choose one without flavour or colour and do not over floss. It can wear a groove in your teeth and tear your gums, making them more susceptible to infection.

Alternatives

Remember that it is not the paste but the brush that cleans your teeth. Equally, it is not how hard you brush, but how long and how thoroughly. Brushing hard and fast will not clean your teeth and may damage your gums. So instead of just running the brush quickly over your teeth and hoping the toothpaste (or mouthwash) will catch what you don't, consider spending at least a minute (if not two) gently but thoroughly brushing your teeth each morning and night.

The type of brush you use is also influential, so for optimum dental health invest in a good brush with lots of filaments packed tightly together. A soft to medium brush is fine for most people (very few really need a firm brush). The angle of the head of the brush is fairly immaterial. Many a weird shaped brush has been developed in recent times – largely to make up for the lazy way in which most people brush. However, there is very little evidence that they are substantially better than a standard brush. Replace your toothbrush regularly, at the first signs of wear.

If you still want to use commercial brands of toothpaste, consider these options:

- *Use less.* In spite of what you see on TV you only need a pea-sized amount of toothpaste (about a quarter of the advertised amount) to clean your teeth. And while children are advised to use a pea-sized amount they can clean their teeth adequately with half this amount, or a 'smear'.

- *Dilute it.* Turn toothpaste into a cream by diluting it with cooled boiled water. Store in a small squeeze-top bottle. It will still foam and clean well.

- *Use a low-fluoride toothpaste.* Among brands aimed at adults, a low-fluoride content is around 500ppm. Several children's brands contain even less and there is nothing (apart from aesthetics) stopping the whole family from using a children's paste to clean their teeth.

- *Buy a fluoride-free brand*, and if you can an SLS-free brand. It will be almost impossible to find a fluoride-free and SLS-free toothpaste in most large supermarket and chemists, though a few, such as Boots and Sainsbury's do offer fluoride-free alternatives. Health food stores will stock a wider range; but check the labels of 'natural' alternatives for detergents and other dubious chemicals.

Very few toothpastes are free from worrying chemicals of one sort or another. If you feel strongly that you should not put anything in your mouth you are not prepared to swallow, then brushing regularly with a tiny amount of bicarbonate of soda or food grade hydrogen peroxide will also clean your teeth adequately. Hydrogen peroxide gives the bonus of gently whitening your teeth.

Since undiluted, straight bicarbonate of soda can be hard on tooth enamel (particularly as you get older and enamel begins to soften) try dissolving a teaspoonful in a little water first and then dipping your tooth brush in the liquid frequently during brushing (when using hydrogen peroxide do the same).

Bicarb does not taste good but can be made more palatable by mixing the dry powder in a small, airtight container with a few drops of peppermint oil. Mix well and store for use as and when you need it.

You can also quite easily make your own toothpaste and mouthwash.

- *Simple toothpaste* can be made from bicarbonate of soda, vegetable glycerine and essential oil of peppermint, lemon or fennel.

  As a general rule use 1 part liquid to two parts bicarbonate of soda; thus 50ml (4 heaped tablespoons) glycerine to 100ml (8 heaped tablespoons) bicarbonate of soda. But you may have to play with the proportions until you get just the consistency you prefer. Add 5 drops of the essential oil of your choice to pep up the flavour. Shake well before use.

This mixture can be stored in a small squeezable or pump dispenser (such as those used for travel) that can be purchased at most large chemists. Use sparingly.

Do not be tempted to make more than about 50ml at a time. Although the essential oils will act as preservatives, when making your own toiletries it is always safest to make small amounts as needed rather than risk the product deteriorating.

Another way of making up this mixture is to substitute a natural mouthwash for the glycerine – one without alcohol, fluoride, chemical preservatives, flavours or colours. Mix this with bicarbonate of soda in approximately the same ratio as above.

- *Whitening toothpaste.* Make up the simple toothpaste and add ½ tsp (2.5ml) of food grade hydrogen peroxide (available at health food stores). Reduce the amount of glycerine you use by this amount.

- *Simple mouthwash.* The simplest mouthwash is a couple of drops of peppermint oil or sage tincture in a cup of water. To make around 100ml of a more complex blend you need:

15ml (1 tblsp) lavender tincture
15ml (1 tblsp) calendula tincture
10ml (2 tsp) aloe juice
30ml (2 tblsp) cooled boiled water
30ml (2 tblsp) vegetable glycerine
5 drops peppermint essential oil

Mix the ingredients together and pour them into a bottle. This mixture will keep for up to 6 months. If you have an infection in the mouth or gums, substitute echinacea, myrrh or golden seal tinctures for the lavender.

Chapter 7

Deodorants and Antiperspirants

Staying Fresh all Day?

Nobody wants to go around smelling like a compost heap (and nobody enjoys being around someone who does), but what price are we paying for keeping shower-fresh all day long?

Deodorants were the first products to be developed to help combat the problem of body odour. They are basically strong perfumes that mask the odour caused by bacteria in your armpits. Later, antiperspirants, which prevent sweat from leaking out of the armpits, were developed. Today there is a huge range of antiperspirants, deodorants and antiperspirant/deodorants on the market in a variety of formulations including creams, roll-ons, solids and sprays. A quick look at the label will tell you that there is not a wide difference between the ingredients used in any of them. Antiperspirant/deodorants are among the most popular choices and typically they will contain:

cyclomethicone, aluminium zirconium tetrachlorhydrex glycine complex, stearyl alcohol, PPG-14 butyl ether, phenyl trimethicone, hydrogenated castor oil, PEG-8 distearate, parfum, gossypium.

Most antiperspirants contain some form of *aluminium*, with the most common forms being aluminium chlorhydrate, aluminium zirconium, aluminium chloride, aluminium sulphate and aluminium phenosulphate.

No one knows exactly how aluminium compounds work to reduce underarm wetness. That in itself is worrying. Aluminium compounds may prevent sweat release by clogging sweat ducts. Clogging the sweat ducts creates pressure from the sweat build-up inside of them and it is thought that this causes the sweat glands to stop secreting.

Other theories say that aluminium perforates the sweat glands so that moisture seeps out into the surrounding tissues rather than coming out through the surface of the skin. Or aluminium may block the transmission of nerve impulses that activate sweat glands.

Either way aluminium is absorbed through the skin, however superficially. The recently acknowledged link between Alzheimer's disease and aluminium has raised a furious debate about whether or not it is safe to put such aluminium compounds into deodorants. One study, in the *Journal of Clinical Epidemiology*, did find a link between Alzheimer's and lifetime deodorant usage. Unfortunately, no other studies have been conducted to confirm the findings (and this is often the case when a study tells us something we do not want to know).

Certainly, aluminium-based deodorants are a major cause of skin irritation and for this reason alone should be approached with caution by consumers. The prolonged use of aluminium zirconium products have been shown to cause granulomas (small nodules of chronically inflamed tissue) under the arms.

Among the other common ingredients in antiperspirant/deodorants are the following:

- *Cyclomethicone* is a volatile oil and *phenyl trimethicone* is a polymer (or type of plastic). Both are derived from silicone, which, in turn, is derived from the silica found in rocks and sand. They are used to help apply the product more smoothly and as skin softeners.

- *PPG-14 butyl ether* is a relative of propylene glycol. It is used as a preservative and solvent. PPG may enhance the skin penetration of other more toxic chemicals. It is poisonous and can be a skin irritant in high concentrations.

- *Hydrogenated castor oil* is castor oil to which hydrogen atom has been added to thicken it. It improves the feel of the product but can be an allergen. *Gossypium* is cotton seed oil. *Stearyl alcohol* derived from animal fats is also used as a lubricating agent to help apply the product more smoothly. It is though to be non-toxic but can cause mild skin irritation.

- *PEG-8 distearate.* Polyethylene glycol (PEG) compounds are derived from natural gas and have many functions in toiletries. Among other things they are moisturisers, emulsifiers, and emollients and antioxidants. Adding PEG to a product will prevent moisture loss during storage. The lower the number the more liquid the product is. PEG compounds can be contaminated with the carcinogen 1,4-dioxane.

- *Parfum* or fragrance based on petrochemicals. Because deodorants work by covering up one smell with another stronger one, the fragrance portion of a deodorant can be very strong. It may be a cause of skin irritation.

Some deodorants also contain *triclosan*, an antibacterial agent that can be absorbed through the skin and that has been shown to cause liver damage in experiment animals. Long-term users are most at risk.

Similarly, some antiperspirants/deodorants have been found to contain dibutylphthalate (DBP), a hormone-disrupting chemical which is implicated in reproductive abnormalities. Pregnant women take note.

Alternatives
If you can, avoid aluminium-based antiperspirants. Aluminium is too toxic and there is too little evidence of the safety of products containing aluminium that are applied to the skin. The presence of ingredients such as *magnesium oxide* and *zinc oxide* will buffer the irritant properties of aluminium- and zirconium-based compounds, but they too can cause skin irritation.

In addition, when selecting antiperspirants and deodorants, you should remember the following:

- *Avoid aerosols* which surround you and those in your immediate area with a cloud of easily inhaled and toxic chemicals. Aerosols can contain planet poisoning HCFCs as well as the neurotoxic and reproductive toxins propane, butane and isopropane.

- *Switch to a solid or stick variety.* Because it is less emollient it is less likely to aid the absorption of ingredients into the skin. Sticks also tend to produce less irritation.

- *Never* apply antiperspirants or deodorants to broken or newly shaved skin. The chemicals they contain will be much more easily absorbed into your system if your skin is damaged in any way.

- *Avoid coloured products.* The colour will not help you stay drier and the coal tar and petrochemical-derived colours used in these products are easily absorbed into the skin and can be carcinogenic.

- Avoid products containing *quaternium 18* which can cause rashes beyond the area of application.

Your health food shop may sell deodorants based on plant extracts and essential oils (but remember to read the label to find out what they really contain). Some also sell *crystal deodorants* (made from mineral salts). These can be very effective but always check what they are made of – some are aluminium-based. Do not buy crystal deodorants whose labels are in any way unclear about the mineral used.

If you are feeling more ambitious you can make your own antiperspirants and deodorants from a few simple ingredients.

- Dust under your arms with *plain cornstarch*. If you don't sweat heavily this may be all you need.

- Equally a simple astringent such as *witch hazel*, which helps to temporarily contract the tissues around the sweat glands, may be all some people need (do not apply to broken skin as it may sting).

- A mixture of unflavoured *vitamin C powder or citric acid and water* works well for some people. Mix 1/4 tsp powdered vitamin C or citric acid in a pint of water. Dab this sparingly under your arms after a bath. Or put it in a spray bottle and spritz a little on. If you cannot wait for it to dry, dust with cornstarch afterwards.

- For a more ambitious spray, the owners of Neal's Yard, one of the UK's most respected natural toiletry manufacturers, recommend the following:

  6 tblsp (90ml) witch hazel
  2 tsp (10ml) vegetable glycerine
  2 drops each of clove, coriander and lavender essential oil
  5 drops each of grapefruit, lime and palmarosa essential oil
  10 drops of lemon essential oil

  Mix the witch hazel and vegetable glycerine together, then add the essential oils. Shake well. Store in a dark glass or plastic bottle with a spray top. The mixture will last for up to 6 months.

- To make an effective *foot deodorant* all you need is 2 tblsp (30ml) witch hazel and 5 drops each of lavender and grapefruit essential oils. Blend the ingredients together, store in a spray bottle (preferably made of dark coloured glass or plastic). Spray regularly on to clean feet. Shake before applying. This mixture will keep for up to 2 months.

- For a *simple foot powder* use cornstarch. If you want a powder that perfumes and deodorises mix 5 drops each of lemon and coriander essential oil in with the cornstarch. Store the perfumed mixture in an old talcum powder dispenser or similar type of container.

Body Sprays

Body sprays serve no useful function. They are not deodorants, they are not quite strong enough to be perfumes and they cannot stop you from sweating. Nevertheless, the market for body sprays, for men and women, has simply exploded in recent years with every manufacturer keen to get a share of the profits.

Advertising for such products suggests that using them will make you more attractive to the opposite sex, or give you the courage to do wild and outrageous things. In reality they are likely to give you a headache (which may stop you from doing wild and outrageous things), cause you to become forgetful, tired and listless and may even make the people around you feel sick as well.

Body sprays are mostly solvents, propellants and fragrances. The perfumes can cause allergic reactions, headaches, dizziness, fatigue and a range of mental symptoms; the solvents and propellants are neurotoxic and have been implicated in reproductive problems such as miscarriage and birth defects.

Alternatives

The best alternative of all is to bin these unnecessary and expensive items. If you must add yet another fragrance to your body try to keep it as natural as possible.

- Use food grade *flower waters*. Rose and orange flower waters can be purchased in your supermarket. While not chemical-free, these have somewhat fewer nasties in them than conventional body sprays. Either splash them on or transfer to a spray bottle for use.

- *You can make a pleasant body spray* from *water and essential oils.* Try mixing 5 drops each of lavender, sage, lemon, rosemary and grapefruit essential oils and 3 drops of peppermint into 10ml vodka. Shake to mix. Then add 100ml white vinegar. Leave this mixture to sit for an hour or so to fix the scent. Add the scented mixture to 500ml of spring or filtered water and shake again.

You can also substitute natural vanilla essence for the essential oils. By learning more about essential oils you can make an infinite number of light splashes and sprays to suit your every mood.

$\wp\supsetneq$

Talc

Talcum powder is one of the mainstays of freshness. We use it liberally on babies' bottoms and to absorb perspiration on hot summer days and nights. A few of us are old enough to remember our mothers having special dishes of talc in the bathroom which had big inviting powder puffs to help you dust your body, and most of the bathroom floor, with the stuff. But time marches on and the romantic illusion of talc has taken a huge knock in recent years.

Talc, or *magnesium silicate*, is made up of finely ground particles of stone. As it originates in the ground, and is a mined product, it can be contaminated with other substances. Asbestos is a good example, and recent reports about the talc used in crayon manufacture being contaminated with this poisonous substance have caused alarm to every parent whose child has ever sucked a crayon.

The harmful effects of talc on human tissue were first recorded in the 1930s. More recently a report from the US National Toxicology Program concluded that talc is carcinogenic. An ominous series of studies has linked talc to ovarian cancer; in these studies talc was observed in a number of ovarian and uterine tumours as well as in ovarian tissue. It has since been confirmed that talc, either placed on the perineum (or on the surface of underwear or sanitary towels), can reach the ovaries via ascent through the fallopian tubes. It is now estimated that women who frequently use talc have three times the risk of developing ovarian cancer compared to non-users.

The talc used in the manufacture of condoms carries a similar risk. In the 1960s the medical journal *The Lancet* reported the first case of a woman who had a significant amount of talc in her peritoneal (abdominal) cavity. Laboratory tests confirmed that the talc in her body matched that found on the surface of her husband's condoms. The authors concluded that talc travelled up through the fallopian tubes and became implanted in her abdomen. Talc sprinkled on diaphragms may also be implicated in such problems.

Talc use is also associated with respiratory problems. Because it is comprised of finely ground stone it can lodge in the lungs and never leave. Babies whose mothers smother them in talc have more breathing difficulties and/or urogenital problems. Women are also at risk since even if they do not use talcum powder on their bodies, they are likely to be using cosmetics (powders, eyeshadows, blushers) that are talc-based.

Alternatives

Do not use products containing talc. Giving up body powders is relatively easy. Giving up your eye shadow may be less so (try applying it with a damp sponge to minimise fallout). But whatever you can do to cut your exposure to talc will benefit your health.

- *Make your own.* You can quickly and easily make a very efficient and inexpensive body powder based on cornstarch. Combine one part baking soda to eight parts of cornstarch. Mix these in a blender and add 10–15 drops of your favourite essential oil (optional). Store in an airtight container (either a jar, or an old talc container, or you can recycle one of those Parmesan cheese shakers).

- *Babies' bottoms do not need talc* or any other powder to stay fresh. Instead let your baby go without nappies as often as possible, or investigate cotton nappies that allow the skin to breathe and have been shown to cause less nappy rash than disposables.

- *Use talc-free condoms,* but beware: many talc-free condoms contain other particles such as vegetable starches, silica (another carcinogen), mica and diatomaceous earth and lycopodium (club moss) spores. Lycopodium can be contaminated with talc, sulphur and/or gypsum and is linked with inflammation of soft tissues. It is not known how many chronic 'women's problems' may be the result of over-use of talc or indeed allergy to the latex used in condoms or other contraceptives such as the diaphragm and all the paraphernalia that goes with them (spermicidal jellies, foams, creams and lubricants).

Feminine Freshness

Feminine deodorants and douches are totally unnecessary. The majority are bought and used simply out of a media-fuelled paranoia which makes women worry that the people around them can detect any faint odours coming from their genital area.

If you really have a problem with strong and unpleasant vaginal odours go to your doctor to sort out the underlying cause of the problem. You may have a low-grade vaginal infection that can easily be cleared up.

Ironically, the use of feminine deodorants can cause vaginal infections that may be the cause of unpleasant odours. More worryingly, in one study involving nearly 700 women, over three years, researchers found that women who used vaginal douches more than once a week experienced a four-fold risk of cervical cancer. It did not matter which preparation was used since all douches alter the chemical balance of the vagina, making the cervix more susceptible to pathologic changes.

Feminine deodorants are almost always aerosols, which means that you inhale harmful chemicals when you use them, and they are always highly perfumed. Douches contain harsh detergents, perfumes and colours none of which should be coming into contact with this delicate area of your body.

Alternatives

Have a little confidence in yourself and toss these toxic products in the bin where they belong. In addition:

- *Bathe daily.* When you wash your genitals do not apply soap directly. It is too harsh and may dry out delicate skin. Use the foam or bubbles from your soap to gently clean yourself.

- *Wear cotton underwear* which allows air to circulate and discourages the bacteria that can cause unpleasant odours.

- *If you are experiencing vaginal itching* or soreness see your doctor and sort out the real problem. It is unlikely that it will be something that you can simply wash away with a douche.

- If you must douche *use a simple mixture of 2 tblsp distilled white vinegar in one pint of water* which will have less of an impact on the vaginal microflora. Use infrequently.

Chapter 8

Facial Care

Facial Cleansers

Your skin tone reflects what is going on inside your body and nothing you put on the outside will be as influential as what you put on the inside. If you ever doubted this take a good long look at yourself in the mirror after a heavy night. A poor diet will also be reflected in the quality of your skin. Similarly, cigarettes and alcohol are highly damaging since both can dehydrate the skin and interfere with its ability to utilise nutrients.

If your skin is looking dull and lumpy it may be working overtime to rid the body of toxins. Women's skin can also change according to where they are in their monthly cycle. There is not much you can do about this, but keeping yourself in good condition through diet and exercise can minimise the impact of normal hormonal changes on your skin.

In spite of the fact that good skin begins on the inside, the skin care market is overflowing with choices that promise to keep you young and wrinkle-free as well as removing dirt, oil and make-up.

A typical creamy facial soap contains:

> sodium cocoyl isethionate, stearic acid, sodium tallowate, aqua, sodium isethionate, coconut acid, sodium stearate, cocamidopropyl betaine, parfum, sodium palm kernelate, sodium chloride, trisodium EDTA, zinc stearate, tetrasodium etidronate, CI 77891.

In this soap, generally considered to be relatively mild, several stand out.

First of all it contains lots of different detergents/surfactants. *Sodium cocoyl isethionate* and *sodium isethionate* are made from entirely synthetic sources and can be irritating to the skin. *Cocamidopropyl betaine* is a highly irritant detergent – one of the chemicals most often implicated in skin irritation. *Sodium tallowate* is synthesised from animal fats and *sodium palm kernelate* is synthesised from the oil of palm kernels. In addition:

- *Trisodium EDTA* and *tetrasodium etidronate* are synthetic preservatives. They are used to stop impurities such as metals from making the product deteriorate. Both can be irritating to the skin.

- Although this bar of facial soap superficially looks white it actually has *colouring* in it to moderate the degree of its white-ness. In this case the white pigment *CI 77891* is titanium dioxide, a known skin irritant and a potential carcinogen.

- *Stearic acid, zinc stearate, sodium stearate* and *coconut acid* are fatty acids added as skin softeners to replace the oils stripped away from the skin by the mix of detergents above.

- *Sodium chloride* is simple table salt, a water softener added to help the product rinse better in hard water.

Liquid face washes are similar in their content to liquid body washes but contain slightly more water. Because liquids are more complex to make than solids, they generally contain more potentially harmful chemicals than detergent bars.

Cleansing lotions are often touted as a better way to clean make-up and grime from your face. Lotions and creams, we are told, can clean without stripping away the skin's natural oils (although products containing solvents will dissolve natural oils, usually replacing them with synthetic ones). The ingredients of most cleansing lotions are nearly indistinguishable from those of facial moisturisers. What is more, even the simplest cleansing lotion can contain a range of suspect chemicals. For example:

> aqua, hydrogenated vegetable oil, glycerine, isopropyl palmitate, polysorbate 60, sorbitan stearate, cetearyl alcohol, panthenol, bisabolol, EDTA, BHT, acrylates/c10–30 alkyl acrylate crosspolymer, methylparaben, sodium hydroxide, 2-bromo-2-nitropropane-1,3-diol.

This product is loaded with preservatives, for instance *Methylparaben* and *2-bromo-2-nitropropane-1,3-diol* (also known as bronopol). Both ingredients can be irritating to the skin. Methylparaben is an endocrine disrupter and over time 2-bromo-2-nitropropane-1,3-diol can break down into formaldehyde – a neurotoxin and potential carcinogen. *EDTA*, or ethylene diamine tetra acetic acid, is a toxic preservative that can cause skin irritation. *BHT*, or butylated hydroxytoulene, is a synthetic antioxidant that can cause allergic reactions. Although a common toiletry ingredient, it is most widely used as an antioxidant in rubber and plastic and in liquid petroleum products such as gasoline and motor oil.

Panthenol and bisabolol are slightly more 'natural' preservatives. Panthenol is a synthetic member of the B-vitamin family. It is used as a preservative and antioxidant for the mix. There is no evidence that it is in any way beneficial to skin. Bisabolol is a plant extract usually derived from camomile. It has antimicrobial properties but can also be a skin irritant.

Among the other ingredients:

- *Polysorbate 60* is a stabiliser and dispersing agent that can be contaminated with the carcinogen 1,4-dioxane.

- *Glycerine*, known technically as glycerol, is a form of alcohol used as a solvent, lubricant and humectant. Used over the longer term, stay-on products with glycerine can cause skin dryness.

- *Hydrogenated vegetable oil* is vegetable oil to which hydrogen has been added in order to thicken it. Hydrogenation is a process that can promote the formation of free-radicals in oils. Free radicals are highly reactive molecules that can damage and age the skin.

- *Sorbitan stearate* is a modified fatty acid derived from beef tallow. It is used as an emulsifier, stabiliser and surfactant. It can cause blackheads in some individuals. *Isopropyl palmitate* is a

---

Antibacterial Cleansers

Teenagers often use antibacterial products, believing that they contain some sort of magic that will wash away spots. They won't. Typically such products contain: aqua, PEG-6, sodium laureth sulphate, glycerine, cocamide DEA, phenoxyisopropanol, triethanolamine, carbomer, C9–11 pareth-8, PEG-75, lanolin, parfum, EDTA, CI 42051.

This example contains three harsh detergent/surfactants, SLES (which can be contaminated with the carcinogen 1,4-dioxane), and the potential hormone-disrupters cocamide DEA and triethanolamine (TEA), as well as the following worrying ingredients:

PEG, or polyethylene glycol, compounds. They are preservatives that can be contaminated with the carcinogen 1,4-dioxane. PEGs can also form carcinogens when mixed with DEA and TEA. *Pareth* is a surfactant that belongs to the same family as polyethylene glycol (PEG).

Lanolin, which comes from wool fat. It is an emollient and thickener. It is non-toxic.

Glycerine. This is a lubricant but can also paradoxically be drying to the skin.

EDTA, or ethylene diamine tetra acetic acid. It is a toxic preservative.

Phenoxyisopropanol, an antibacterial agent. It is manufactured by combining carcinogenic phenol (coal tar) with the solvent isopropanol. It is an irritant and allergic reactions are possible.

Carbomer, a gelling agent that can be irritating to skin and eyes.

Parfum. It does not clean the face and can be a source of skin and airway irritation.

Colour – in this case *CI 42051*, or Patent Blue V. It does not clean the skin and is added only to appeal to the eye.

If you have spots it is a result of hormones and poor lifestyle. Plenty of sleep, water in favour of sodas, cutting down on sugar and fat, investigating food or other allergies (if the problem is really severe) and a little patience is the best way to tackle them.

chemically modified fatty acid. While considered relatively non-toxic its use may increase the skin's ability to absorb more toxic chemicals such as carcinogens.

- *Cetearyl alcohol*, also known as cetyl stearyl alcohol or emulsifying wax, can be animal, vegetable or petrochemical in origin. It is a thickener and emulsifier with no known toxicity.

- *Acrylates/c10–30 alkyl acrylate crosspolymer* are emulsifiers and surfactants that help the product stick to the skin and can make the skin feel softer. They are not thought to be toxic.

- *Sodium hydroxide*, also known as lye or caustic soda, is the same chemical used in oven and drain cleaners. In small amounts it can be used to adjust the pH of the product. However, it can also be irritating to the skin.

And this product was advertised as being kind to sensitive skin!

Alternatives

Do not use special cleansers for you face. These are not much more than a mixture of diluted dish detergent and oil. Instead:

- Try to *avoid liquid cleansers* which are more expensive and have more harsh chemicals in them.

- If you want to use conventional bars *opt for glycerine soaps* which are among the mildest.

- *Switch from detergent bars to castile or pure vegetable oil soaps.* These will clean your face without stripping it completely of natural, beneficial and protective oils (and you will not need to invest in separate bars for body and face).

- Whatever you use remember to *rinse well*. Film left behind by neglectful rinsing can irritate skin.

- *Forget about antibacterial cleansers.* Simple soap and warm water are among the best antibacterials and the most effective ways to keep your face clean. They are also less expensive and unlikely to contain as many suspect chemicals.

- *To remove make-up* and city grime use almond or jojoba oil on a cotton ball, then rinse with warm water. Follow with a cool rinse or mild astringent such as witch hazel.

Moisturisers

While not strictly used to clean the face, moisturisers are an integral part of many women's routine. The name is misleading since 'moisturiser' would seem to indicate that it will put water into the skin. No moisturiser, no matter how expensive or how many ingredients it contains, can do this. Even if your moisturiser could push water back into your skin, it would be quickly disseminated throughout the rest of your body, rather than held there for an indefinite period.

Generally, moisturisers work by stripping away natural oils and replacing them with synthetic ones. They also put a thin resinous film over the skin to temporarily stop moisture loss. Additionally, some also use humectants – chemicals that act like water magnets – drawing moisture from the surrounding air and holding it close to the skin.

Moisturisers typically contain:

> aqua, octyl methoxycinnamate, isohexadecane, zinc oxide, glycerine, tocopherol acetate, sucrose, polycottonseedate, dimethicone copolyol, steareth-21, steareth-2, cetyl alcohol, stearyl alcohol, behenyl alcohol, c13–14 isoparaffin, cyclomethicone, polyacrylamide, laureth-7, DEA-oleth-3 phosphate, DMDM hydantoin, disodium EDTA, parfum, iodopropylyl butylcarbamate, BHA, BHT

- *Octyl methoxycinnamate* and *zinc oxide* are sunscreens. These ingredients are very popular in facial moisturisers. Both can cause skin irritation.

- *Glycerine*, known technically as glycerol, can be of animal or vegetable origin. It is a form of alcohol used as a solvent and humectant (water magnet). Stay-on products with glycerine should be avoided if possible since glycerine (in common with other humectants such as PEGs) draws moisture from the closest, most abundant, source. If you live and/or work in a dry environment the closest source of moisture will be your skin. Used daily, glycerine products will draw moisture from your skin, not keep it in.

- *Tocopherol acetate* is a synthetic Vitamin E, added as an antioxidant. There is no evidence that it can help improve skin condition.

- *Sucrose* is sugar or sugar cane extract. It is an alpha hydroxy acid (AHA). AHAs can irritate and are implicated in increased photosensitivity of the skin.

- *Polycottonseedate* is an emollient derived from cotton seed oil.

- *Dimethicone copolyol* is a more waterproof form of dimethicone that sticks to skin and hair better. It is a polymer based on silicone, used as a conditioner in hair care products and as a skin protectant. It is not considered toxic, but with prolonged use it can make the skin look dull. *Cyclomethicone* is another silicone-based emollient and solvent, used primarily to improve the feel of a product.

- *Steareth-21* and *Steareth-2* are waxy compounds used as emulsifiers. They are part of a larger group of ethoxylated alcohols that are toxic and potentially carginogenic in their own right and may be contaminated with the carcinogen 1,4-dioxane.

- *Cetyl alcohol* is not really an alcohol but an emollient and emulsifier; it is a waxy substance closely resembling human sebum. It is often used in the production of synthetic sebum. It can cause hives. *Stearyl alcohol* is often used as a lubricating agent to help apply the product more smoothly. Appears to be non-toxic but can cause mild irritation to the skin. *Behenyl alcohol* is also fatty alcohol. It has antioxidant properties and can irritate the skin.

- *C13–14 isoparaffin* is a solvent and lubricating agent derived from petrochemicals. It is a relative of mineral oil used to apply the product smoothly, but may cause skin irritation and increased photosensitivity of the skin. *Isohexadecane* is also an isoparaffin.

- *Polyacrylamide* is used to improve the feel of the skin by putting a thin layer of plastic-like material on top of it.

- *Laureth-7.* Laureth compounds are emulsifiers made from combining ethylene oxide with lauryl. They can be contaminated with the carcinogen 1,4-dioxane.

- *DEA-oleth-3 phosphate* is a surfactant and relative of polyethylene glycol (PEG). It is produced by ethoxylation and can be carcinogenic in itself as well as being contaminated with the carcinogen 1,4-dioxane.

- *DMDM hydantoin*, or diemethylol dimethyl hydantoin, is a water-soluble preservative that can act as a formaldehyde-releasing agent. Should not be combined with DEA as this can cause carcinogenic NDELA to form. It can be toxic at high levels.

- *BHA*, or butylated hydroxyanisole, is a synthetic preservative found to be carcinogenic in some animals. *BHT*, or butylated hydroxytoulene, is another synthetic preservative that can cause allergic reactions.

- *Disodium EDTA* is a preservative used to keep impurities such as metals from causing the mixture to degrade. It is not generally considered toxic externally, although sensitisation is possible.

- *Iodopropylyl butylcarbamate* is a preservative and fungicide. It is a skin irritant and mutagen (able to cause cellular mutations).

Long-term use of moisturisers on the face (and on the body) may actually suffocate the skin, causing more harm than good. There is also some evidence that moisturisers can make the skin more susceptible to damage caused by synthetic detergents used in many facial and body care products.

If your moisturiser includes fruit acids, or AHAs, it may cause premature ageing of your skin as well as increased susceptibility to damage from the sun's ultra violet rays.

Alternatives
If you are going to continue to use commercial products choose those with the fewest ingredients and watch out in particular for those that may be contaminated with carcinogens. It is the oil and wax content of moisturisers that holds moisture next to the skin, so why not consider simple vegetable oils to maintain the skin's suppleness? They will do the same job at a fraction of the price and you will have the advantage of actually knowing what you are putting on your face. As a general rule, use lighter oils such as apricot kernel, coconut or jojoba oil for normal skins and heavier oils such as avocado and evening primrose oil for older or drier skins. Rosehip oil is also considered a rich and nourishing oil for the face.

More Stuff for your Face
The amount of stuff we put on our faces is simply staggering and since facial skin is thinner than skin elsewhere, the potential for absorbing toxic chemicals is much greater. While most skincare products promise to help us look younger, prolonged use can dry and age the skin considerably. It may not show when you start using it at age 24, but by the time you are 54 the damage will be noticeable and mostly irreversible.

Exfoliating scrubs typically contain harsh detergents, emulsifiers and abrasives. An abrasive can be anything from ground fruit pips, to talc, to more worrying particles such as aluminium oxide. Others contain skin drying agents such as alcohol. A simple face cloth will do the job just as well so it is a bit silly to invest in anything else. Exfoliants are a waste of money, may unnecessarily damage your skin through over-enthusiastic use, and will certainly add to your body's toxic burden.

Toners are now accepted as the necessary intermediate step between cleaning and moisturising. In fact this step is an invention of the marketing world. Clean skin is as toned as it needs to be.

Some toners use alcohol to dry the skin and make it feel tighter. Others promise to close your pores, thus making your skin look younger and firmer. Unfortunately, your pore size is genetically determined and pores do not open and close. If they did your face would wobble like a jelly throughout the day. What toner does is put astringent chemicals on your face that lower the skin temperature. This affects the underlying tissue which will temporarily contract, giving that characteristic 'tight' feeling that you get after using a toner.

You can get the same feeling standing in a cold shower or splashing cold water on your face, so why waste your money on dubious chemical mixes? If you absolutely need to buy something to 'tone' your skin, buy simple products like distilled witch hazel or try using a weak solution of cider vinegar in demineralised water.

Shaving Cream

Wet shaving with a blade is one of the oldest ways of removing unwanted hair (from any part of the body). There is now a wide variety of shaving creams or foams on the market for both men and women. These creams and foams look nice, they feel nice and some even smell nice. But they can contain some not-so-nice ingredients, for example:

> aqua, plamitic acid, triethanolamine, laureth-23, isobutane, aloe barbadensis, paraffinum liquidum, stearyl alcohol, cetyl dimethicone copolyol, propane, lauramide DEA, parfum, PEG-150 distearate, BHT, imidazolidinyl urea, methylparaben, propylparaben, quaternium-15.

- *Triethanolamine* (TEA) and *lauramide DEA* are detergent/surfactants that can be irritating to the skin. More worrying, when mixed with formaldehyde-forming PEG compounds (see below) they can contain carcinogenic nitrosamines.

- *PEG-150 distearate.* PEG compounds are derived from polyethylene glycol. They are used as stabilisers, humectants and moisturisers. The lower the number the more liquid the product is. PEGs are formaldehyde-forming agents so should not be used in products containing DEA, MEA or TEA. PEGs can also be contaminated with the carcinogen 1,4-dioxane. Humectants attract water from the most abundant source. In dry atmospheres leave-on products containing PEGs will draw moisture from the skin.

- *BHT* and *imidazolidinyl urea* are preservatives. BHT, or butylated hydroxytoulene, is a synthetic antioxidant that can cause allergic reactions. Imidazolidinyl urea is the most commonly used preservative after the parabens group. It can release formaldehyde into the formulation and is most dangerous when used in combination with ethanolamines (DEA, MEA, TEA). It can also cause allergic/sensitivity reactions.

- *Parabens* or parahydroxybenzoic acid esters are also synthetic preservatives. They are easily absorbed and can be irritating to the skin. They are also estrogen mimics. Listed in order from the most estrogenic to the least, the parabens group includes: butylparaben, propylparaben, ethylparaben, methylparaben.

- *Quaternium-15* is a conditioner and preservative that can be irritating to the skin. It is also a formaldehyde-forming chemical that produces nitrosamines when used with DEA and TEA.

- *Paraffinum liquidum* is mineral oil, a by-product of petroleum refining. In shaving cream it is used as a lubricant. Mineral oil can cause skin irritation and there is evidence that it can be carcinogenic and may increase skin photosensitivity.

- *Isobutane* and *propane* are propellants. They are neurotoxic (damaging to the central nervous system) in high concentrations. They also contribute to the phenomenon of global warming.

- *Stearyl alcohol* is often used as a lubricating agent to help apply the product more smoothly. It appears to be non-toxic but can cause mild irritation to the skin. *Plamitic acid* is a fatty acid added as a surfactant and skin softener.

- *Cetyl dimethicone copolyol* is a silicone derivative used as a conditioner and as a skin protectant. It is not considered toxic.

- *Laureth-23*. Laureth compounds are emulsifiers. Laureth-23 can be contaminated with the carcinogen 1,4-dioxane.

- *Aloe barbadensis*. Aloe extract is added to soothe the face. Many shaving foams use plant extracts. These are generally non-toxic but can cause allergic and sensitivity reactions in some individuals.

The example above did not list parfum on the label, even though it was strongly fragranced. The majority of shaving creams will contain fragrances and some will contain totally unnecessary colours that add to the toxic burden of the product.

Alternatives
A wet shave should be a *wet* shave. Getting hair thoroughly wet before shaving means you may require less cream or foam. Modern

shaving creams are loaded with lubricants because most men and women skip the essential wetting part of the shave. In addition:

- *Use a shaving soap.* It will not be a soap really (unless you buy it at a specialist natural healthcare shop), but a detergent bar. However, you will be able to avoid the problems associated with solvents and propellants found in shaving foams. Use a brush to get lots of foam unless you have very dry skin. The shaving brush may irritate dry skin so soap up on your hands first before applying the foam to your skin.

- *Use shaving oil.* More and more companies are making these nice oils which usually contain added essential oils to smooth and soothe the surface of the face while shaving. Opt for vegetable oil bases in favour of mineral oils.

- Buy an *electric shaver.* The shave will not usually be so close, but it avoids a lot of unnecessary exposure to harsh chemicals.

Chapter 9

Hair Care

Shampoos

How many different types of shampoo have you tried in your lifetime? And how often have they fallen short of what they promised to do? If you have had more than your share of shampoo failures it could be because there are only a limited number of cleansing agents considered suitable for use in hair cleaning products. If you are unsure of the truth of this statement, compare expensive designer brands with their cheaper cousins. Often the only genuine difference between them is the price.

The main function of a shampoo is to clean the hair. Its function is so simple that advertisers have to work doubly hard to make it sound more complicated and exciting than it actually is. Thus using a particular brand brings with it the promise of harmony, lust for life, nourishment and adoration by members of the opposite sex. Some shampoos are apparently so remarkable that they not only clean your hair but give you an orgasm as well! However, underneath the puffery, a shampoo is just a bottle of detergent.

The word shampoo is derived from 'chapo', a Hindi word meaning to massage or knead. The first shampoos were simple solutions of soap and were invented by British hairdressers during the heyday of the Empire. Modern shampoos, however, are usually a mixture of several different detergents and surfactants, typically sodium lauryl sulphate, ammonium lauryl sulphate, monoethanolamine (MEA) lauryl sulphate, diethanolamine (DEA) lauryl sulphate, and triethanolamine (TEA) lauryl sulphate.

Generally the strongest detergent is used in the greatest measure. Then milder detergents/surfactants, which modify the harshness of the first detergent, are added. These other detergents/surfactants can also add foaming ability and conditioning properties. At this point you may wonder why manufacturers do not just start with a single mild and effective detergent. It is a good question – why not write to the manufacturers and ask them? If you get a straight answer let me know!

For the formulator, the choice of detergent for any particular shampoo is as much a matter of aesthetics as it is cleaning. For instance, sodium lauryl sulphate is not very soluble in cold water and so it cannot be used to make shampoos that look 'clear'. For these shampoos other compounds such as ammonium lauryl sulphate or TEA lauryl sulphate are used. Some shampoos contain extra ingredients to make them produce more foam even though foam does not actually clean your hair.

---

pH balanced?

There is that term again. All shampoos – whether they claim to be or not – are pH balanced. But do not be fooled into thinking this is for your benefit.

Many personal care products need to have their pH controlled during manufacture or the mixture will fail. Get the pH wrong and foams would not foam, gels would collapse and those nice pearlescent liquids would separate into unsightly lumps.

Carcinogens in the Mix

As detailed in Chapter 1, a number of common ingredients in shampoos such as 2-bromo-2-nitropropane-1,3-diol (bronopol or BNP), DMDM hydantoin, diazolidinyl urea, imidazolindinyl urea and quaternium 15 break down into formaldehyde during storage. When formaldehyde-forming agents mix with amines, e.g. diethanolamine (DEA), triethanolamine (TEA) and monoethanolamine (MEA), they form carcinogenic N-nitroso-diethanolamine, or NDELA.

Nitrosamine formation is particularly problematical in shampoos since we use them so frequently and in such great quantities. It is estimated that when you wash your hair with a shampoo contaminated with NDELA your body absorbs more carcinogenic nitrites than if you had eaten a pound of bacon.

A further problem is that products containing the milder laureth detergents (as in sodium *laureth* sulphate and ammonium *laureth* sulphate) can be contaminated with the carcinogen 1,4-dioxane. Laureth compounds are part of a larger family of chemicals called ethoxylated alcohols. Many of these, including polyethylene glycol (PEG), polyethylene, polyoxyethylene, oxynol and other *eth* chemicals (as in laur*eth*) can be contaminated with 1,4-dioxane. Products containing polysorbates 60 and 80 can also be contaminated with this chemical.

The contamination of the raw materials used to create sodium laureth sulphate was noted as far back as 1978 and has been confirmed in a recent review. Yet little has been done to address this issue. As if all this was not enough, the latest published research reveals that the preservatives methylchloroisothiazolinone and methylisothiazolinone (which together are sometimes called Kathon) have the potential to cause skin cancer. None of these ingredients is exclusive to shampoos. You will find them in nearly all your toiletries as well as in household cleaners.

As manufacturers fall over themselves to make a more 'scientific' and 'improved' shampoo, the list of chemical ingredients grows. So today a typical mild shampoo will contain:

aqua, sodium laureth sulphate, cocamide DEA, glycerine, lecithin, caprylic/capric triglyceride, tocopherol acetate, retinyl palmitate, stearomine, alcohol denat, panthenol, polytrimonium hydrolysed soy protein, polytrimonium lysine, polytrimonium vegetable amino acids, cocamidopropyl betaine, cocamidopropylamine oxide, cocamido-propyl hydroxysultaine, propylene glycol, glycol distearate, cocamide MEA, laureth-10, formic acid, sodium chloride, parfum, methylchloro-isothiazolinone, methylisothiazolinone, magnesium nitrate, magnesium chloride, citric acid, triethanolamine, hexylene glycol.

This bottle of shampoo has 31 listed ingredients including several detergents and surfactants. These include *sodium laureth sulphate*

(SLES), a mild cleaning agent but one that can be contaminated with the carcinogen 1,4-dioxane. *Cocamide DEA* and *cocamide MEA* are strong detergents and *triethanolamine* (TEA) is a surfactant. These can react with formaldehyde-forming agents to form nitrosamines. DEA, TEA and MEA are also hormone-disrupting chemicals. *Cocamidopropyl betaine* is a detergent that is a strong allergen and skin irritant. *Cocamidopropylamine oxide* is a non-ionic surfactant used to make the liquid more viscose and reduce static. *Cocamidopropyl hydroxysultaine* is a surfactant and conditioner. It can cause skin irritation in some individuals.

Along with these it contains the following:

- *Propylene glycol* (PG), used as a preservative, solvent, fixative, skin conditioning agent and surfactant. It is the most common moisture-carrying vehicle in toiletries after water. PG is poisonous, irritating to the skin and may enhance the skin penetration of other more toxic chemicals.

- *Methylchloroisothiazolinone* and *methylisothiazolinone*, synthetic preservatives that can cause skin irritation and may raise your risk of skin cancer.

- *Laureth-10*, an emulsifier. It is an ethoxylated alcohol that can be contaminated with the carcinogen 1,4-dioxane.

Harsh detergents can strip natural, protective oils from the hair. To counter this effect several conditioning agents are added to the mix. These include:

- *Polytrimonium hydrolysed soy protein, polytrimonium lysine* and *polytrimonium vegetable amino acids*, amino acids used to repair damaged hair.

- *Caprilic/capric triglyceride*, an oily liquid from plants, vegetable oils and dairy fats. It is used as a barrier agent, emollient and solvent in a wide variety of cosmetics.

- *Hexylene glycol*, used as a humectant, plasticiser, solvent and emulsifier. Can cause skin irritation in some individuals. Vapours can be irritating to the eyes and lungs. Repeated oral and dermal applications have been shown to affect the kidneys and liver.

- *Glycerine*, known technically as glycerol; is a form of alcohol used as a solvent and humectant.

- *Formic acid*, a solvent and plasticiser. It is added to coat the hair to make it feel softer. It is toxic by ingestion and inhalation and can be a severe skin and eye irritant.

- *Magnesium chloride*, used as a hair softener. It is toxic by ingestion and is a known human mutagen (able to cause mutations in cells).

Inevitably there are also preservatives including:

- *Citric acid* which, like all AHAs, can cause irritation. AHAs have recently been listed as a major cause of adverse reactions to soaps, shampoos and moisturisers.

- *Alcohol denat* or denatured alcohol, used as a preservative. It is also a solvent that is dangerous to inhale and easily turns to a toxic vapour in a hot shower.

- *Lecithin*, an antioxidant with low toxicity to humans. It is also used as an emulsifier to hold the mixture together.

- *Panthenol, tocopherol acetate and retinyl palmitate*, synthetic vitamins added as antioxidants to keep the product from deteriorating. There is no evidence that they 'nourish' the hair.

Finally *sodium chloride*, or table salt. This thickens the mixture and is also a water softener added to help the product rinse better in hard water. *Parfum* is derived from petrochemical sources. It irritates the lungs, respiratory tract, skin and eyes.

Using nice hot water to shampoo your hair actually increases the rate of absorption of these chemicals into your body, so if you do not like cold showers, choose your shampoos carefully.

Alternatives
Hair care begins with what you eat, not what you wash with. Hair is 95 per cent protein so if your hair is limp consider whether your protein intake is adequate.

Hair needs to be thoroughly wet before shampooing. This helps to spread the shampoo evenly throughout the hair. For really clean hair you need to rinse thoroughly. In the 'wash and go' culture this is a step that most of us rush through. In addition:

- *Read the label.* It behoves you to find a shampoo with the fewest possible ingredients in order to limit your exposure to toxic chemicals. Do not buy products that contain formal-dehyde-forming agents and amines.

- *Use less.* Sounds obvious, but most of us use far too much shampoo to get the job done. A single shampoo using half what you normally use will clean your hair perfectly well. Always tip your head well back when rinsing to avoid any getting on to your eyelids and into your eyes.

It is not easy to make a shampoo (as we have come to know it) based on natural ingredients. At some point you will have to add detergent. But you can make mixes that are less concentrated and that expose you to fewer harmful chemicals in each wash. So:

- *Dilute it.* Mix your shampoo with an equal amount of water and put this mixture in an old and well-rinsed shampoo bottle. Adding a little table salt (appx. 1 tsp per 100mls of liquid) will help to thicken the mixture.

  For a variation on the diluting theme, try this: Half fill an old shampoo bottle with your regular shampoo. Top up with an equal amount of water or a strong herbal infusion (such as camomile) mixed with 1tsp of coconut oil (this comes as a solid but can be melted in the hot infusion). Olive or jojoba oils are also good choices. Add 1–2tsp of table salt to thicken or alternatively 1–2tsp of sodium bicarbonate to help soften the water and aid rinsing (dissolve this in the water/tea mixture *before* mixing). This mixture will still foam well and will also clean and condition your hair. Always give it a shake before using, as the ingredients in some shampoo mixtures will separate when the mix is altered.

What's in Dandruff Shampoos?

Dandruff shampoos are made with detergents to which anti-flaking agents, such as coal tar, zinc pyrithizone, salicylic acid and selenium sulphide are added. While they can relieve itching and decrease flaking, no dandruff shampoo can control dandruff completely.

Sulphur and salicylic acid work by breaking the flakes into smaller less noticeable pieces. It is thought that coal tar, selenium sulphide and zinc pyrithizone can slow the production of flakes. Beyond this there is little known about how, exactly, anti-dandruff shampoos work.

Of all the anti-flaking agents zinc pyrithizone and coal tar are considered to be the most effective in controlling dandruff.

All anti-flaking agents have some side effects. They can be irritating to both skin and eyes. In particular salicylic acid, an ingredient of aspirin, can be severely irritating and is a poison if swallowed. Coal tar is a known carcinogen and can be an irritant when inhaled or when it comes into contact with skin.

Dandruff is caused by a fungus. It is most effectively treated from within. Effective dietary measures include cutting out sugar and yeasty foods, supplementing with B-complex and pro-biotics (acidophilous and bifidobacterium) and making sure you drink plenty of water each day.

Externally, a more natural alternative is to make an effective antidandruff lotion with 1tsp each of rosemary and thyme essential oils mixed in to 100 ml (3½ fl oz) of apple juice and 2tblsp (30ml) cider vinegar. Apply this at bedtime or on days when you can let your hair dry naturally to help keep dandruff at bay.

- *Use a castile soap.* These come in both bar and liquid form. Because of their high oil content, castile soaps will wash and condition your hair. Alternatively use a pure vegetable oil soap (these can usually only be found at specialist suppliers). These types of soaps can also be used on the body, thus saving you money as well as being safer.

Repairing the Damage with Conditioners

Have you ever noticed how shampoo bottles always recommend that you use a conditioner afterwards? This is because the detergents used in shampoos are often so harsh that you need to use a conditioner to repair some of the damage done by their use. Healthy hair rarely needs conditioning. Hair damaged by detergent use (and other assaults) almost always does.

Conditioners do not repair hair. They cannot penetrate the hair shaft and make it stronger. Instead they coat the hair with chemicals that temporarily glue down the damaged hair shaft – giving the illusion of healthier hair.

All shampoos, no matter how 'mild', will strip away the protective layer of sebum (natural oils produced by the scalp) that coats your hair. Stripping the sebum away exposes the outside layer of the hair, known as the cuticle. The cuticle is made up of translucent over-lapping cells that are arranged like the shingles on a roof. When these cells are disturbed they can rub against each other and become damaged, resulting in the social horror of flyaway hair. Stripping away the sebum also leaves the cuticle (the inner part of the hair shaft) vulnerable to damage.

Many shampoos contain conditioning agents that smooth down the cuticle and cover it with a synthetic version of sebum. As the following example shows, conditioners contain many of the same ingredients as shampoos – just in different proportion and without the foaming agents:

aqua, cetearyl alcohol, quaternium-18, stearamidopropyl dimethylamine, glycerine, quaternium-80, lecithin, caprylic/capric triglyceride, tocopherol acetate, retinyl palmitate, stearomine, alcohol denat, panthenol, polytrimonium hydrolysed soy protein, polytrimonium lysine, polytrimonium vegetable amino acids, hydroxyethylcellulose, dimethicone, PVP, oleamine oxide, parfum, steareth-21, glyceryl stearate, citric acid, DMDM hydantoin.

This revitalising conditioner, which is a companion to the shampoo example above, has 25 listed ingredients. Many of these are emulsifiers (help to hold water and oil together in the mix) and emollients (prevent moisture loss from the hair) that are fairly safe to use. For instance, *cetearyl alcohol*, also known as cetyl stearyl alcohol or emulsifying wax, can be of animal, vegetable or petro-chemical origin. It is a thickener and emulsifier with low toxicity.

Likewise *lecithin* is an emulsifier and antioxidant derived from soya oil with low toxicity to humans. *Steareths* are emulsifiers and are part of a larger group of ethoxylated alcohols that are toxic and potentially carcinogenic in their own right and may be contaminated with the carcinogen 1,4-dioxane. *Glyceryl stearate* is a white, waxy solid emulsifier and moisturising agent based on animal fats. *Hydroxyethylcellulose* is a chemically modified form of cellulose used as a thickening agent. *Oleamine oxide* is a non-ionic surfactant that aids the interaction between the conditioners, the water and the hair.

In addition the product contains many preservatives:

* *DMDM hydantoin*, or diemethylol dimethyl hydantoin, is a water-soluble preservative that can act as a formaldehyde-releasing agent. It can be toxic at high levels.

* *Quaternium-18 and Quaternium-80* are preservatives and surfactants that can cause skin irritation. Quaternay ammonium compounds are added to both conditioners and shampoos to allow easy detangling during combing – they are the same

compounds as those used in fabric softeners. Quaternay compounds can aid the absorption of other more toxic chemicals into the skin and can cause skin irritation.

- *Citric acid* is a preservative. It is also an alpha hydroxy acid that can cause mild skin irritation. Some people react more violently than others to AHAs.

- *Panthenol, tocopherol acetate and retinyl palmitate* are synthetic vitamins added as antioxidants to keep the product from deteriorating. They do not 'nourish' hair.

- *PVP*, or polyvinylpyrrolidone, is a petrochemical-based polymer added to leave a smooth, resinous, water-resistant coating on the hair and to aid styling. It is the same basic ingredient that keeps soda bottles sturdy but pliable. PVP is associated with liver damage and cancer in animal studies.

- *Caprilic/capric triglyceride* is an oily liquid from plants, vegetable oils and dairy fats. It is used as a barrier agent, emollient and solvent in a wide variety of cosmetics.

- *Dimethicone* is oil derived from silicone (which is derived from silica found in rocks and sand). It is used to help apply the product more smoothly, and as an emollient and hair softener.

- *Polytrimonium hydrolysed soy protein, polytrimonium lysine* and *polytrimonium vegetable amino acids* are oils and amino acids used to coat damaged hair and give the appearance of a smooth texture.

- *Glycerine*, or glycerol, is a form of alcohol used as a solvent, humectant and moisturiser.

- *Alcohol denat* or denatured alcohol, is a preservative and solvent that is dangerous to inhale and easily turns to a toxic vapour in a hot shower.

- *Stearamidopropyl dimethylamine* is part of a larger group of glycol ethers (the same family as PPG and PEG compounds). Glycol ethers are solvents and wetting agents that are quickly absorbed into the skin. In the body they are reproductive toxins.

- *Parfum* derived from petrochemicals can be a significant source of skin and eye irritation. The perfume used in conditioners is even stronger than that used in shampoos since it is meant to linger on the hair.

Alternatives

The most sensible alternative is not to use a conditioner at all. You can see from the examples above that many shampoos (even the ones that are not specifically 'conditioning' shampoos) contain some conditioning ingredients. If you follow the advice for using or making milder shampoos you should not need a conditioner.

If you are going to continue to use conventional conditioners, once again the best advice is to use less. You can do this in two ways. First only use a conditioner once or twice a week. Second, you can put less on your hair at each washing. This can be hard to do straight from the bottle so try diluting it instead. To do this, half fill a *well-rinsed* conditioner bottle with regular conditioner and then top up with water. Shake before use.

Alternatively, consider some more basic alternatives:

- *Regular trimming* and keeping the use of hair-destroying stuff like rollers, curling irons, hairdryers and lots of styling gels to a minimum will also help keep hair looking good.

- *Condition before you wash.* Contrary to what we have been encouraged to do, this is the best way to keep hair soft and manageable. Rubbing some good quality oil through your hair a half hour or so before you shampoo, or better still at night before bed, will provide the same conditioning action as the complex and hazardous ingredients in your usual conditioner.

Good choices for conditioning oils include olive oil for deep conditioning and coconut, jojoba or almond oil for light to medium conditioning. Add 2 tblsp of honey to 100ml (3¹/2 oz) oil to help 'repair' split ends. Leave on ¹/2 hour before shampooing.

• After shampooing *rinse with cider vinegar* (which is pH neutral) to remove detergent deposits and lightly condition the hair.

Sprays, Gels and Other Hair Goop

Most hair-styling products are used to make up for poor quality cuts and years of abusive practices such as harsh shampoos, hair dryers, heated rollers and styling wands.

As with shampoos, many of us tend to collect half-used pots, tubes and bottles of this stuff – a good indication that they never quite do what they are supposed to. Apart from poor performance, most styling products contain pretty dubious chemicals that your hair, skin and lungs would be better off without.

Hair Spray

Hair spray is essentially plastic dissolved in a solvent and put in a pressurised can or pump spray. It works by gluing strands of hair together so that they form a stronger structure that can then hold a style. Recently it has been reported that hair spray also contains phthalates – hormone-disrupting chemicals that are used to keep plastics and vinyl soft and pliable.

Many women find it hard to believe that hair spray is just liquid plastic. If you are one of them try this test. Spray your usual hair spray on to your bathroom mirror and leave it to dry. If you spray it in a thick enough layer you should be able to peel it off in a single sheet. But even if you do not you will be able to scrape off little shavings of plastic – the same stuff that you deposit on your hair and into your lungs each time you use hair spray.

Perhaps unsurprisingly, there is a medical condition known as hairdresser's lung – a respiratory disease caused by chronic exposure to hair spray. Even though the average consumer is unlikely to get this disease, using hair spray regularly can do other nasty things to your health. Your nose is lined with tiny hairs that filter out dirt from the air you breathe. When hair spray gets into your nose, these little hairs become sticky and begin to attract dust and pollution until they become saturated with dirt – at which point they stop filtering pollutants effectively.

Other distressing side effects of hair spray use include nail abnormalities. When you spray and then style your hair, using your fingers, the spray is deposited on the nail where it can cause the new nail to grow poorly or predispose it to infection. Breathing difficulties and contact dermatitis after hair spray use are also common.

The ingredients that cause such problems are typically:

> alcohol denat, dimethyl ether, VA/vinyl butyl benzoate/crotonates copolymer, aminomethyl propanol, cyclopentasiloxane, dimethicone copolyol, PPG-3 methyl ether, parfum, CI 11223/1.

- *Alcohol denat* is a solvent that is dangerous to inhale. It is made from mixing ethlyl alcohol (ethanol – the same stuff found in beers, wines and spirits) with a chemical such as benzene to make it poisonous. It is harmful by ingestion, inhalation and skin absorption and can irritate skin, nose and throat.

- *Aminomethyl propanol* (AMP) is another alcohol that can irritate skin and eyes. It is also a formaldehyde-forming agent, which means you may be breathing in formaldehyde with each spray.

- *Dimethyl ether*, a relative of propylene glycol, is a solvent that is easily absorbed through the skin. Once in the body it can bioaccumulate and is known to be a reproductive toxin. *PPG-3 methyl ether* is also a propylene glycol compound. It is a solvent and fixative and part of a larger family of glycol ethers that are known to be reproductive toxins. It is easily absorbed through the skin and through inhalation.

- *VA/vinyl butyl benzoate/crotonates copolymer* is a petroleum-based vinyl acetate that forms a thin plastic-like film on the hair and aids styling. It can cause mild skin irritation.

- *Dimethicone copolyol* is a more waterproof form of dimethicone that sticks to skin and hair better. It is a polymer based on silicone and used as a conditioner in hair care products. It has a tendency to build up on the hair shaft, eventually making the hair appear dull and is not, therefore, currently favoured. It is not considered toxic. *Cyclopentasiloxane* is a lubricant, also based on silicone.

- *Parfum* is petrochemically-derived. Many of the ingredients in hair spray do not smell very nice, which is why strong perfumes are often used.

- *CI 11223/1* are azo colours that are known carcinogens. Why a product in a metal container should have any pigment is a mystery.

Hair Gel

Like hair spray, hair gel is a type of resin or plastic. Hair gels are generally more concentrated than hair sprays, which is why you can use them to make your hair stick in so many different gravity-defying shapes. Typically they contain:

aqua, PVP, PVP/dimethyl-aminoethylmethacrylate copolymer, laureth-23, triisopropanolamine, carbomer, methylparaben, methicone copolymer, parfum, propylparaben, DMDM hydantoin, polyquaternium-4, hydrolysed wheat protein, hydrolysed wheat starch, malvia sylvestras, rosa canina, melilotus officinalis, propylene glycol, disodium EDTA, diasolidinyl urea, phenoxyethanol, ethylparaben, butylparaben, potassium sorbate, CL 61570.

This example is an herbal 'extra hold' styling gel. But apart from the addition of a few plant extracts, the content is much the same as many others – a combination of plastics and preservatives.

- *PVP*, or polyvinylpyrrolidone, is a petrochemical-based plasticiser and is the same basic ingredient that keeps a soda bottle sturdy but pliable. It leaves a smooth, water-resistant coating on the hair that aids styling. PVP has been associated with liver damage and cancer in animal studies. *PVP/dimethyl-aminoethylmethacrylate copolymer* is a mixture of PVP and silicone-derived polymers. It performs a similar function to PVP and carries the same risks. *Methicone copolymer* is derived from silicone and used as a hair conditioner and styling agent.

- *Carbomer* is a gelling agent. It is a synthetic polymer used to thicken, stabilise and promote the shelf life of the product. It can be irritating to skin and eyes.

- *Parabens* or parahydroxybenzoic acid esters are synthetic preservatives that can be derived from plant or petroleum sources. They are easily absorbed by and irritating to the skin and have exhibited hormone-disrupting activity. In order of the most estrogenic to the least, the parabens group are: butylparaben, propylparaben, ethylparaben, methylparaben. This product contains all four.

- *Phenoxyethanol* is a preservative often used in combination with parabens. It is manufactured by combining carcinogenic phenol (coal tar) with ethylene oxide. It is an irritant and allergic reactions are possible.

- *Propylene glycol* (PG) is used as a preservative, solvent, fixative, skin conditioning agent and surfactant. It is a poisonous skin and eye irritant. PG also enhances the skin penetration of other more toxic chemicals. *Diasolidinyl urea* is another preservative that can cause dermatitis. It is also a formaldehyde-forming agent. *Disodium EDTA* is a preservative and chelating agent (used to isolate impurities such as heavy metals in the mixture and keep them from causing the mixture to collapse). It is not generally considered toxic externally, although allergic reactions are possible. *DMDM hydantoin*, or diemethylol dimethyl hydantoin, is a water-soluble preservative that can act as a formaldehyde-releasing agent. It can be toxic at high levels.

- *Laureth-23* is part of a family of emulsifiers made from ethylene oxide. Laureth compounds can be contaminated with the carcinogen 1,4-dioxane.

- *Triisopropanolamine (TIPA)* is an alkanolamine (same family as DEA, TEA and MEA). It is a hormone-disrupting solvent and surfactant.

- *Polyquaternium-4* is a quaternay ammonium compound used as a surfactant, conditioner, thickener and emollient.

- *Hydrolysed wheat protein* is wheat protein broken down chemically to remove hydrogen. It is used as a conditioner for skin and hair. *Hydrolysed wheat starch* is wheat starch that has gone through the same process. It is also used to improve the feel of skin and hair.

- *Parfum* is from petrochemical sources.

- *Malvia sylvestras (mallow flower)*, *rosa canina (rosehip powder)*, *melilotus officinalis (clover)* are plant extracts. In large quantities they can be used for their conditioning properties. However, they are unlikely to be included in enough quantity to be therapeutic.

- *CL 61570*, or D&C green 5, is a synthetic colour for which there is no safety information.

Alternatives

Apart from not using it, there are really very few alternatives to things like hair spray and gel. The best alternative is to stop buying hair styling products and use the money you save on a really good haircut. With a really good haircut you should not need to apply lots of glue to your hair to keep it in place. Do this and you may find that you and your hair live happily ever after.

Try reserving the use of sprays and gels for when you really need them, like on special occasions when you want your hair to stay in place. In addition, when purchasing hair sprays, gels and other styling agents:

- *Read the label.* Try to buy products with the fewest and least toxic ingredients.

- *Use pump sprays* instead of aerosols. You are still at risk of inhaling the noxious chemicals in the mixture, but you will be avoiding inhaling toxic propellants as well – a small step in the right direction.

- *Make your own simple hairspray.* While no home-made spray will keep your hair in place as effectively as the liquid plastic of commercial brands, you can make a gently holding spray from the following ingredients. (Note also that home-made hairspray will not dry as quickly as conventional sprays.)

  1 lemon
  300ml (10fl oz) filtered or spring water
  2 tblsp vodka

  Chop the lemon into pieces and place in a saucepan with the water. Bring to the boil, cover and simmer gently for 20 minutes (until the lemon is tender). Strain, pressing hard on the lemon to extract as much liquid as possible. When completely cool, put into a spray bottle and add the vodka (to prolong the shelf life). This will keep for 1 month.

- *Make your own simple gel.* You can make a simple light styling gel from unflavoured gelatine (preferably vegetable based). Mix 2 tsp (or $1/2$ of half a pack) of gelatine powder with $1/2$ pint of hot water or herbal infusion into which you have melted 2 tsp of coconut oil. Stir well. When the mixture has cooled slightly add 10 drops of your favourite essential oil (light oils like neroli, geranium, lavender are good choices). When this sets it can be used just like a regular setting gel. The coconut oil (which is a good hair conditioner) will harden into small flecks in the mix but as you rub it in your hands before use it will quickly melt again. Store in the refrigerator and use within 1 month.

- *To make a light spray gel*, adapt the recipe above as follows. Dilute 100g of the set gel with a further 100ml of warm water (to melt the oil) and 3 tblsp vodka (tequila is also good, but any clear alcohol will do). The alcohol will act as a preservative (nevertheless you should not keep it for more than one month) and solvent (to dissolve the oil). You can add 10 drops of your favourite essential oil if you wish to have a strongly scented mix. Put this in a spray bottle and use to provide light styling power and to revamp limp curls.

Hair to Die For

Both human and animal studies show that the body rapidly absorbs the carcinogens and other chemicals in permanent and semi-permanent dyes through the skin during the more than 30 minutes that dyes remain on the scalp. So if you use permanent, semi-permanent, shampoo-in or temporary hair colours you are increasing your risk of developing cancer.

Problems with hair dyes were first noted in the late 1970s when several studies found links between the use of hair dyes and breast cancer. In 1976 one study reported that 87 of 100 breast cancer patients had been long-term hair dye users.

In 1979, another study found a significant relationship between the frequency and duration of hair dye use and breast cancer. Women who started dyeing their hair at age 20 had twice the risk of those who started at age 40. Those at greatest risk were the 50 to 79-year-olds who had been dyeing their hair for years, suggesting that the cancer takes years to develop.

A year later another study found that women who dye their hair to change its colour, rather than masking greyness, were at a threefold risk of developing breast cancer.

Research continued and in the early 1990s Japanese and Finnish studies also linked hair dye use with breast cancer. More recently, a jointly funded American Cancer Society and FDA study found a four-fold increase in relatively uncommon cancers, including non-Hodgkin's lymphoma and multiple myeloma, among hair dye users.

As if this wasn't enough, a study in the February 2001 edition of the *International Journal of Cancer* found a link between long-term hair dye use and an increased incidence of bladder cancer.

As a general rule, the darker the shade of the dye, the higher the risk of breast cancer; thus women who use black, dark brown or red dyes are at the greatest risk.

In fairness there are problems with studies into hair dye and cancer risk. Some involve small numbers of women working in the cosmetic industry. These women are exposed to the known carcinogens in hair dyes – diaminotoulene, diaminoanisole, and phenylenediamines, coal tar dyes, the dioxane found in detergents, solvents, nitrosamines and formaldehyde-releasing preservatives – in much greater concentration than the rest of us. However, taken together they point towards an increased overall risk that may be unacceptable for some.

There are also studies that dispute the cancer risk of hair dyes. Nevertheless, it is still believed that long-term hair dye use may account for as many as 20 per cent of all cases of non-Hodgkin's lymphoma (the cancer that killed Jackie Kennedy – a long-term black hair dye user) in women.

Although diaminotoulene and diaminoanisole were removed from hair dye products some 20 years ago it is likely that past use of dyes containing these chemicals is a cause of some cases of breast cancer today.

Today the main risks for future breast cancers come from colours, solvents, preservatives and phenylenediamines and it is probably no coincidence that hair dye manufacturers now cover themselves by including a warning on their product labels alerting women to the presence of *phenylenediamines* and *ammonia,* as well as *napthol* and *resorcinol* (both of which are phenols; see listing under toilet cleaners for more).

Other hair dye ingredients such as *chlorides* are highly irritating to the mucous membranes. Chloride fumes can irritate the lungs and eyes and cause burns or rashes on the skin. Hair colorants also contain several ingredients known to aid the absorption of other toxic chemicals into the bloodstream. These can include *propylene glycol, polyethylene glycol,* fatty acids such as *oleic, palmitic* and *lauric acid* and *isopropyl alcohol,* to name but a few.

Alternatives
If you intend to keep dyeing your hair consider the safer options:

* *Read the label.* If you dye your hair use the safer alternatives that are currently on the market. These should not contain phenylenediamines (though many so-called natural hair colours do). Never buy products that are in any way unclear about their ingredients.

* *Read the label again.* This time look for dyes. Avoid products that use colours like Acid Orange 87, Solvent Brown 44, Acid Blue 168 and Acid Violet 73 – these are also carcinogenic.

* *Do not dye your hair too often.* Leave the maximum amount of time between applications.

- Leave hair dyes on the head for the *minimum required time.*

- Hair colorants made entirely from *plant based ingredients* are the safest choice. However, these are few and far between. Pure herbal hair dyes will need to be left in the hair significantly longer then synthetic dyes, but have the advantage of conditioning the hair while they colour.

- *Go natural.* In a world of lookalike bleach blondes and unnaturally red redheads you will probably be the standout.

Chapter 10

Skin Care

Body Lotions

Like facial moisturisers, body lotions make use of humectants to draw moisture to the skin and synthetic oil barriers to stop moisture from evaporating out of the skin. While most humectants and oils are relatively harmless, their effect is, at best, temporary. More worrying, however, is the fact that several of the ingredients commonly used in body lotions can make the skin more permeable, allowing more toxic ingredients to be absorbed into the body.

The ingredients of a typical body lotion are not much different from those of a hair conditioner. And basically the two products perform the same function – replacing natural oils with synthetic ones and then coating the body with a thin waterproof layer. Typically a body lotion might contain:

> aqua, paraffinum liquidum, propylene glycol, isopropyl palmitate, stearic acid, prunus dulcis, glyceryl stearate, cetyl alcohol, triethanolamine, polysorbate-80, glyceryl stearate SE, carbomer, dimethicone, sorbitan oleate, hydrolysed collagen, foeniculum vulgare, humulus lupulus, mellissa officinalis, chamomilla recutita, achillea millefolium, viscum album, parfum, quaternium-15, trisodium EDTA, BHA, methylparaben, propylparaben.

The label proclaims that this product is 'lanolin free', but watch out for what it does have in it.

- *Paraffinum liquidum*, or mineral oil, is a by-product of petroleum refining. It contains no minerals and cannot 'nourish' the skin. It is not well absorbed by the skin and instead puts a barrier on top of it, blocking pores and preventing the skin from breathing or excreting. Paraffinum can cause skin irritation and photosensitivity and is potentially carcinogenic.

- *Triethanolamine*, or TEA, is a surfactant that can react with formaldehyde-forming agents to form carcinogenic nitrosamines. It is also an endocrine disrupter.

- *Polysorbate*-80 is a dispersing agent and stabiliser that can cause mild skin irritation. Some polysorbates are contaminated with the carcinogen 1,4-dioxane.

- *Propylene glycol* (PG) is a preservative, solvent, fixative, skin conditioning agent and surfactant. It is a poisonous, irritating agent that can enhance the skin penetration of other more toxic chemicals.

- *Quaternium-15* is a synthetic cationic surfactant and a formaldehyde-forming preservative. Mixed with TEA it can form carcinogenic nitrosamines. It can also aid the absorption of other more toxic chemicals into the skin and can cause allergic sensitivity reactions.

- *Isopropyl palmitate* is a chemically modified fatty acid. While considered relatively non-toxic its use may increase the skin's ability to absorb more toxic chemicals such as carcinogens present in the mixture.

- *Stearic acid* is a fatty acid that can be derived from either animal or vegetable fats. Its primary function is as a skin softener. It has a low toxicity but skin irritation is possible.

- *Glyceryl stearate* is a by-product of detergents based on animal fats. It is a white, waxy solid used as an emulsifier and moisturising agent. Likewise, *cetyl alcohol* is not really an alcohol but an emollient and emulsifier. It is a waxy substance closely resembling human sebum and is often used in the production of synthetic sebum. Cetyl alcohol can cause hives in some people.

- *Carbomer* is a synthetic gelling agent used to thicken, stabilise and promote the shelf life of the product. It can be irritating to skin and eyes.

- *Dimethicone* is oil derived from silicone (which is derived from silica found in rocks and sand). It is used as a skin softener and to help apply the product more smoothly.

- *Sorbitan oleate* is an emulsifier and surfactant synthesised from olive oil.

- *Hydrolysed collagen* is usually derived from the skin of young animals, though some is vegetable derived. Collagen is the protein that makes up the fibrous support system from which skin is made. While the hydrolysation process breaks down the collagen chemically, it is still composed of molecules too large to be absorbed through the skin. Instead they simply put a smooth coating on top of the skin.

- *Trisodium EDTA* is a preservative used in detergent bars and other toiletries, particularly those that contain metals. It removes trace metals from the mix to keep them from reacting with other ingredients and causing the mix to degrade. It can be irritating to the skin and mucous membranes.

- *BHA* or butylated hydroxyanisole, is a synthetic preservative, found to be carcinogenic in some animals.

- *Methylparaben* and *propylparaben* are parahydroxybenzoic acid esters – synthetic preservatives that can be derived from plant or petroleum sources. They are irritating to the skin. They are also easily absorbed by the skin and have exhibited hormone-disrupting activity.

This particular lotion also has synthetic fragrance (*parfum*) and several plant extracts including *prunus dulcis* (almond milk), and herbs that can be added either as powders or oils *foeniculum vulgare* (fennel), *humulus lupulus* (hops), *mellissa officinalis* (lemon balm), *chamomilla recutita* (camomile), *achillea millefolium* (yarrow), *viscum album* (mistletoe).

Body Oils

Most body oils are based on mineral oil. Some, like baby oil, are 100 per cent mineral oil, with added perfume.

Mineral oil is manufactured from crude oil. Because it comes from petrochemical origins it can cause sensitivity reactions. Sensitivity to petrochemicals can show itself over time as headaches but also, according to allergists, more serious disorders such as arthritis and diabetes. Petrolatum, paraffin or paraffin oil and propylene glycol are all forms of mineral oil.

While it is used for its lubricant qualities which in the short-term appear to make the skin softer, used over the longer-term mineral oil can make the skin dry out. This is because mineral oil dissolves the skin's natural oils, thereby increasing water loss (dehydration) from the skin. Mineral oils may also increase the skin's sensitivity to sunlight and have been linked to an increased risk of skin cancer.

Hand and Nail Creams

Like body oils and lotions, hand and nail creams are essentially mineral-oil based. There is little difference between most body lotions and those that are supposed to be specifically for the hands, although hand creams tend to have even more emollients in them. Hands become dry and cracked because they are more exposed to the elements and more in use than almost any other part of the body. You also wash your hands more often and expose them to detergents and other cleaning products more often each day.

While not as thin as facial skin, the skin on your hands is still thin enough to allow the penetration of noxious chemicals. To keep them soft, and to avoid absorbing toxic chemicals into you body, always use gloves when doing cleaning jobs, and substitute vegetable oils and vegetable oil-based creams for petroleum-based ones.

Among women who switch toiletry brands or try new products, mineral oils have been shown to be the major cause of new skin irritation including rashes and spots. There is no good reason for this kind of suffering. Vegetable oils are better absorbed by the skin, and they soften and moisturise without any of the risks associated with petroleum by-products.

Sun Creams

Sun exposure and the use of sunscreens is one of the most complex and contradictory areas of skin health. On the one hand, exposure to the fresh air and sun is vital for a healthy body. Sunlight, for example, is an important source of vitamin D, necessary for the development and maintenance of bones and teeth. But at the same time, too much sun exposure can raise your risk of skin cancer.

Relying on sunscreens as your sole means of protection is fraught with problems since the protection they offer is never guaranteed. In addition, most commercial sun creams contain a mixture of harsh and harmful chemicals that present their own risks. For example, some of the chemicals in sunscreens are thought to cause disruption or permanent damage to the nervous, immune and respiratory systems. Young children may be especially susceptible to sunscreen chemicals and their toxic side-effects. Among the most harmful are benzophenones, which can cause allergic reactions, and PABAs which have been shown to form carcinogenic nitrosamines when mixed with other chemicals.

The effectiveness of any sun cream depends on its UV absorption, its concentration, formulation and ability to withstand swimming or sweating. As a general rule, the higher the sun protection factor (SPF) the greater the number of chemicals in a sun lotion or cream. It is not uncommon for sun creams to contain three or more sunscreen agents as well as perfumes, insect repellants and a host of other chemicals besides. Though many of the ingredients used in sunscreens have been tested individually, studies of the long-term effects of combinations of sunscreen agents, applied liberally over an extended period of time, are rare.

There are two main types of ultraviolet rays: UVA and UVB. The SPF factor in your sun cream is for UVB protection only. Most UVB rays are filtered out by the ozone layer. Those that do get through stimulate the skin's pigment to produce melanin, our natural defence against sunlight. UVA rays are not filtered out by the ozone layer and penetrate the skin at a deeper level, so they have the potential to cause more skin damage. Gauging UVA protection is a little more difficult, though most creams now put UVA information on their labels as well.

Some experts believe that, when exposed to UV rays, sunscreens like oxybenzone can break down into chemicals that destroy or inhibit the skin's natural defences against sunlight. This leaves it vulnerable to the free radicals produced by exposure to sunlight. Free radicals are the toxic by-products of metabolism. Free-radical damage to the skin is implicated in skin cancer, premature ageing and other damage to the skin.

Similarly, sunscreens such as padimate-O are thought to absorb harmful UV rays. But, as scientists point out, once absorbed this energy still has to be released somewhere, usually directly on to the skin where it is metabolised into free radicals that can actually increase the risk of skin cancer.

What is in your sun cream?
Two basic types of sun cream are available on the market today: chemical sunscreens, which act by absorbing ultraviolet light, and chemical sunblocks, which reflect or scatter light in the visible spectrum and UV spectrum. Both types are associated with skin irritation.

These are the most common chemical *sunscreens:*

Benzophenones, common skin sensitisers that can provoke allergic reactions in some individuals. Common benzophenones include oxybenzone, dioxybenzone and sulisbenzone.

PABAs are formaldehyde-forming chemicals that can form carcinogenic nitrosamines when combined with amines such as DEA, TEA and MEA in the mixture. PABAs can cause skin irritation. Common PABAs are: p-aminobenzoic acid, ethyl dihydroxypropyl PABA, padimate-O (ocyl dimethyl PABA), padimate A and glyceryl PABA.

Cinnamates are common skin irritants. This group of chemicals includes: cinoxate, ethylhexyl-p-methoxycinnamate, octocrylene and ocytl methoxycinnamate.

Salicylates are also skin irritants and are associated with a high rate of dermatitis among users. Those commonly used in sun creams include: ethylhexyl salicylate, homosalate, octyl salicylate and neo-homosalate.

Natural Sunscreens?

Read labels for natural and 'organic' sunscreens carefully. Usually they are simply the same old ingredients with added plant extracts and oils. Can the addition of these natural ingredients really prevent sunburn? No, of course not. For example:

Aqua – water; does not prevent sunburn.

Glycerin – a lubricant used in moisturisers to make them feel good and apply more smoothly. It has no sun-blocking ability but can dry the skin, making it more vulnerable to sun damage.

Octyl palmitate – a relative of Vitamin C; there is no evidence that it can provide protection from the sun.

Retinyl palmitate – also known as pro-Vitamin A or pro-retinol; there is no evidence of sun protection.

Tocopherol acetate – a relative of Vitamin E; again, there is no evidence that it has any effect as a sunscreen.

Other ingredients including aloe vera, carrot oil, chamomile, borage oil and avocado oil are used as fillers, stabilisers or preservatives. They are seldom present in high enough quantities to protect or nourish your skin and none of them has proven sun-blocking ability.

Other common sunscreen agents include methyl anthranilate, digalloyl trioleate and avobenzone (butyl methoxy-dibenzoyl-methane).

The most commonly used chemical *sunblocks* are: zinc oxide, titanium dioxide and red petrolatum.

Bear in mind that in addition to sunscreens, sun creams also inevitably include all the same ingredients as body lotions such as mineral and other synthetic oils, PEGs, TEA and other surfactants, preservatives and fragrances.

Alternatives

Recent research in the *British Medical Journal* suggests that individuals who use sunscreens may actually be at an increased risk of developing skin cancer. This is because high SPF creams give sun worshippers a false sense of security, encouraging them to venture out during peak periods and to stay out in the sun much longer than would normally be considered safe.

So what is a sun lover to do? The only safe recommendation is not to rely on sunscreens as your sole method of protection. There is no doubt that sunscreens can be useful, but they should not be applied over large parts of the body for an extended period of time.

Around 80 per cent of our total lifetime exposure to the sun comes during childhood. So it is especially important to make sure children have some protection from strong sunlight. If you want to teach your children to be safe in the sun, lead by example. Research into children's voluntary use of sun creams suggests that they follow their parents' lead when it comes to covering up and slopping on the cream. Other useful things you can do include the following:

- *Keep babies under 6 months old out of the sun.* Baby sun creams with a high SPF probably have the greatest number of toxic chemicals and are not suitable for the delicate and permeable skin of babies.

- *Limit time in the sun.* It is probably best to avoid being out when the sun is at its strongest, between the hours of 11 am and 2 pm.

- *Cover up.* When you and your children are in the sun, make sure that you are covered up as much as possible. Consider investing in special sun protective clothing that stops UV rays from penetrating to the skin.

- *Cover up some more.* Buy hats and sunglasses and wear them. Encourage your children to wear them too.

- *Go under cover.* If you are going to be out on the beach for a long time (and especially if you have children) use a special protective tent or umbrella on the beach so that you have somewhere to sit out of the sun.

Hair Today, Gone Tomorrow

Depilatories, which come in gels, creams, lotions, aerosols and roll-ons, are chemical razor-blades. Usually they contain a highly alkaline chemical such as calcium thioglycolate that dissolves the protein structure of the hair, causing it to separate from the skin surface. The problem is that skin and hair are similar in their composition. What damages one can also damage the other. For this reason, if you are going to use depilatory creams it is particularly important to follow the directions, not leave them on for too long and not re-apply too regularly.

PART 3

HARMFUL HOUSEHOLD PRODUCTS

Chapter 11

Air Fresheners

Freshen or Poison?

In spite of their name, air fresheners do not freshen the air. Instead they are among the most concentrated sources of poisons and pollution in the home. While the accepted hype is that air fresheners remove unwanted odours and replace them with clean fresh smells, most do no such thing.

Air fresheners work in several ways. They can contain nerve-deadening agents designed to interfere with your ability to detect smells or they can coat the nasal passages with a thin, undetectable oily film that also prevents odour detection. Alternatively, they can cover up one smell with another, more powerful one. Rarely do they actually remove or break down unpleasant odours.

Years ago air fresheners were something that you sprayed into a room once in a while for effect. Today you can plug them in to an electric socket and have a continual release of strong perfume throughout your home. But the trend towards using continuous room perfumes means that your body gets no break from exposure to these chemicals and we simply do not know enough about the long-term effects of this exposure to deem it safe (see chapter 3 for more).

Air fresheners also provide a good example of how manufacturers leap on current trends and use them for profit. For instance, many air fresheners today promise 'aromatherapy' for your home. It is all nonsense and even if your air freshener does not make you feel sick in the short term, it may add to longer-term problems.

Animal experiments, for example, suggest that the chemicals in air fresheners may weaken the body's defences by making the skin more permeable. Allergic reactions to these products can range from sneezing and watery eyes to full-scale asthmatic attacks. Aerosol air fresheners contain neurotoxic propellants and can be harmful to the lungs if inhaled in high concentrations or over prolonged periods of time. Solids and plug-ins produce the same symptoms and can also be poisonous to children and pets if ingested.

While air fresheners are often marketed as luxury items – you have a perfume, your husband has aftershave, doesn't your home deserve its own scent? – their ingredients are far from romantic. Typically they contain:

> water, para-dichlorobenzene (PDCB), naphthalene, formaldehyde, sodium bisulphate, glycol ethers, ethanol and other solvents, various propellants such as HCFCs (hydrochlorofluorocarbons), butane, propane, and isobutane, disinfectants such as cresol, preservatives (e.g. quaternay ammonium salts), and petrochemical-based fragrances.

- *Para-dichlorobenzene (PDCB)* is irritating to skin, eyes and throat and has been shown to cause liver damage in animal studies. It is commonly used as a moth repellant and general insecticide, as well as a mildew control agent. Symptoms of exposure to PDCB fumes include drowsiness, weakness, headache, swollen eyes, stuffy head, loss of appetite, nausea, vomiting, and throat and eye irritation.

- *Naphthalene* is poisonous to humans and a potential carcinogen. It is one of a number of possible petroleum distillates that may be present in air fresheners and is also a pesticide, insecticide and fungicide. It is also used in the manufacture of lacquers and varnishes. Like all petroleum distillates it is irritating to the skin, eyes, mucous membranes and to the upper respiratory tract. Exposure can cause nausea, vomiting and profuse sweating.

- *Formaldehyde* is a preservative and suspected human carcinogen that (among other things) alters your sense of smell and can cause respiratory irritation. Anyone with asthma, lung infections or similar ailments can be severely affected by formaldehyde. It can also cause stuffy nose and itchy or watery eyes, nausea, headache, and fatigue. Some other preservatives used in household products, such as quaternium compounds, can also break down into formaldehyde.

- *Sodium bisulphate*, or sulphuric acid, is a strong deodoriser that is corrosive and irritating to the skin and eyes. If ingested it can be damaging to body tissues. Its fumes are a common trigger of asthmatic attacks.

- *Cresol*, a relative of phenol, is a disinfectant that attacks the central nervous system, the respiratory system, liver, kidneys, pancreas, spleen, skin and eyes. It can enter the body via the lungs or the skin. Repeated or prolonged exposure to low concentrations of cresol can produce chronic systemic poisoning. Symptoms of poisoning include vomiting, difficulty in swallowing, diarrhoea, loss of appetite, headache, fainting, dizziness, mental disturbance and skin rash.

- *Solvents.* The solvents used in household products are called 'organic' – not because they are free from harmful chemicals but because they contain the carbon atoms common to all life on earth. All organic solvents are hazardous. They are easily absorbed through the skin and via their fumes. Solvents have been associated with liver and kidney problems, birth defects and nervous system disorders. Most recently, solvent exposure has been linked to an increased risk of developing Parkinson's disease. Many of these problems develop slowly over time and are thought to be the result of chronic low level exposure to solvents such as ethyl or isopropyl alcohol, propylene glycol and ethanol.

- *Propellants* are found in all aerosol sprays. They emit fine particles of spray that can be easily and deeply inhaled into the lungs and quickly absorbed into the bloodstream. Chemicals that may be relatively harmless on your skin (because their molecules are too large to be absorbed through the skin) may become extremely dangerous if inhaled as a mist. Common propellants include HCFCs (hydrochlorofluorocarbons), propane, butane and isobutane. Propellants are associated with nervous system disorders, skin, eye, throat and lung irritation, lung inflammation and liver damage. They are also highly flammable.

• *Perfumes* derived from synthetic sources. Such perfumes may be combinations of hundreds of undisclosed chemicals. Many synthetic perfumes are associated with central nervous system problems, with allergic and asthmatic reactions, and with headaches, nausea and lethargy.

Britain is the biggest producer and user of aerosols in Europe, with the average household buying 36 cans a year. Again manufacturers argue that there is probably too little of any one chemical in each formulation to do harm. But recently in one of the largest surveys of its kind, involving 14,0000 pregnant women, researchers concluded that frequent use of household aerosol sprays could be making women and their babies ill.

The study, reported in *New Scientist* magazine in 1999, advised 'caution' over the use of aerosols and air fresheners more than once a week. It found evidence that women who used aerosols and air fresheners most days suffered a quarter more headaches than those who used them less than once a week. There was an increase of 19 per cent in postnatal depression associated with women who frequently used air fresheners.

The study also found that babies under six months old frequently exposed to air fresheners had 30 per cent more ear infections than those exposed less than once a week. Babies frequently exposed to aerosols were also more likely to suffer from diarrhoea.

The researchers noted that the chemicals present in many aerosols, such as xylene, ketones and aldehydes, had already been associated with so-called 'sick building syndrome' and that we now may be looking at a similar syndrome in homes.

Fresh Air?

In the 4 September 1996 edition of the medical journal *Medical Monitor*, Dr Richard Lawson described his experience with patients exposed to room fresheners.

He described a female patient who was ordinarily 'cheerful and open'. In every other way her life – marriage, family, work, finances – was happy. Yet she had come to him complaining of heart palpitations and insomnia and feeling dizzy and 'muzzy'.

Initially, the doctor did what many of his colleagues would have done – sent her off with a prescription for tranquillisers. But the drug did nothing to relieve her symptoms. On a subsequent visit to his office the woman said that she noticed she became very much worse in the bathroom. She then produced a circular plastic air freshener out of her bag, saying she had been fine since she took it out of the house.

A week later, another young women consulted the doctor with textbook symptoms of 'nervous breakdown'. Recently she had found herself wandering outside at night and she was experiencing weepiness, tremors, anxiety and feelings of unreality. Remembering his recent experience, the doctor asked about her use of air fresheners. The woman revealed that her mother had just placed one in her room three weeks earlier. Within nine days of removing the air freshener her symptoms disappeared.

Dr Lawson recorded a further 50 cases of what is usually written off as 'anxiety/hyperventilation syndrome' which were 'cured' by removing 'fresh air' perfumes from the house.

Alternatives

You do not need air fresheners and the best alternative is not to buy them at all. Simply keeping your home clean and well ventilated solves most odour problems. If your house is really smelly then you need to get to the root of the problem – often animals, blocked drains, damp conditions, gas from your cooker or heater, off-gassing from new carpets, furniture or wall paper, over-full bins and cigarette smoke – rather than mask it.

To remove unwanted smells consider some of these measures:

- *Buy pump sprays instead of aerosols.* But do not kid yourself that these are much healthier. You will still be breathing in alcohols and perfumes as well as a toxic mix of other chemicals. Pump sprays just mean that you avoid breathing in propellants – a baby step in the right direction.

- *Open a window.* Fresh air is the best air freshener there is. To save money and environmental resources we have all become accustomed to living and working in sealed premises. Unfortunately, sealing in the heat also seals in every smell and every chemical and gas that may be present in the environment. Unless you live in a dreadfully polluted space, opening a couple of windows to allow cross-ventilation is always the best way to get rid of a bad smell. Even in winter this is unlikely to raise your heating bill significantly.

- *Buy houseplants* such as English ivy, spider plants, peace lilies and philodendrons; these can help remove unpleasant odours and gases from your environment (see information in chapter 4).

- *Make your own spray.* A simple air freshener can be made by combining equal amounts of water and white vinegar in a spray bottle. Add 20–30 drops of your favourite essential oil (peppermint, lemon, pine and geranium are all good choices) and shake before using. As both vinegar and essential oils can be eye irritants do not put your face in the area you have just sprayed and do not spray directly into your face.

- *Make you own room freshener* with a 200g box of baking soda and 10–15 drops of your favourite essential oils. Mix these well and place in a cardboard box (you can paint or decorate this) or a small dish. This is a fairly effective way of cutting down on everyday household odours.

  Another tip is to place a few cotton balls sprinkled with vanilla extract in an open dish or bowl and leave these to freshen the room. This is only a short-term solution and after a day or so the vanilla will dry up and stop working. Nevertheless, it is useful for after parties to cut through smoky smells and freshen small spaces such as closets and cars.

- *Use potpourri with caution.* Most commercially prepared potpourris use petrochemical-derived fragrances. Look at the label and if it says something like 'a blend of natural aromas', or 'natural' anything, it is probably not natural at all but 'nature identical' (sees chapter 3). If you want to avoid these, make your own potpourri or buy an unscented variety and scent it with your own essential oils rather than with the synthetic potpourri oil blends. These will not last as long but they will be less harmful to your health.

Be careful when buying aromatherapy candles, also marketed as 'natural' alternatives to conventional air fresheners. Most use petrochemical fragrances instead of natural oils. Secondly, the candle itself is usually made from wax derived from petrochemicals. As this burns it can release toxic soot into the atmosphere.

Which Essential Oils?

The oils you use in homemade room sprays can be tailored to your needs. For everyday freshness any oil you enjoy is fine. For a sick room oils which freshen and have an antibacterial effect are best. So try a combination of orange and cinnamon or cloves. Bergamot, eucalyptus, pine, grapefruit, tea-tree and lemon are all good at battling bacteria. Try using them in the kitchen or bathroom as well to freshen and keep bacteria at bay.

Chapter 12

All-Purpose Cleaners

Not Nice or Clean

While it may seem that there is a huge selection of different household cleaners on the supermarket shelves, each with its own special 'formulation', most household cleaners contain the same range of ingredients.

While all-purpose cleaners may have to deal with tougher grime than toiletries, they work in essentially the same way to loosen grease and grime and facilitate a clean rinse. In order for these products to live up to the promise of effortless cleaning they tend to be much stronger and more concentrated than necessary for everyday use. Nevertheless, few of them are significantly more efficient than simple soap and hot water for loosening everyday dirt and grime from household surfaces.

The ingredients in all-purpose cleaners are a combination of synthetic detergents, grease-cutting agents such as solvents, bleaches and disinfectants. Depending on the mix of ingredients these cleaners have been shown to be irritating, in varying degrees, to skin, eyes, nose and throat and can be corrosive if swallowed. Those that contain small amounts of phosphates are an environmental hazard.

Liquid Cleaners

All-purpose cleaners come in liquid, cream and powder forms. The typical liquid cleaner will be made up of:

detergents/surfactants, solvents such as ethanol, isopropanol, butyl cellosolve or glycols, disinfectants such as ammonia, bleach (sodium hypochlorite), phenol and pine oil, fragrances and colours. Abrasive or cream formulas contain fine particles of abrasives such as plastic, silica, calcite, feldspar and quartz. Most also contain undisclosed preservatives such as quaternay ammonium compounds and EDTA.

- *Detergent/surfactants* can contain individual chemicals that are known to be carcinogens or they may contain a mixture of chemicals that interact with other chemicals to form carcinogens. Surfactants such as diethanolamine (DEA) and triethanolamine (TEA) and morpholine can react with undisclosed formaldehyde-forming preservatives or other contaminants to form carcinogenic nitrosamines that are easily absorbed into the skin. Some detergent/surfactants can also be contaminated with the carcinogen 1,4-dioxane.

- *EDTA.* As an alternative to phosphates, some products contain chemicals like ethylene diamine tetra acetic acid (EDTA) to reduce water hardness. EDTA stabilises the bleach and foaming agents in detergent products, preventing them from becoming active before they are immersed in water. While the health effects of some chemicals, e.g. headaches and skin rashes, are direct, others are indirect. EDTA, for example, can be irritating to the skin but it also binds with toxic metals in the environment and remobilises them, carrying them back into our drinking water supplies and food, especially fish and shellfish.

- *Solvents* enter the body through the skin and lungs. They are hazardous chemicals associated with liver and kidney problems, birth defects and nervous system disorders. In addition, many solvents adversely affect the central nervous system, producing symptoms of drunkenness. *Butyl cellosolve*, for example, is a neurotoxin (a chemical that attacks the nervous system) that can quickly be absorbed into the skin, which is why you should always use rubber gloves when using cleaning fluids. Inhalation can cause headaches and nausea and damage to internal tissues and organs such as the liver and kidneys. Some all-purpose cleaners can contain *ethylene glycol monobutyl acetate*, a relative of butyl cellosolve, which can cause damage to internal organs through skin absorption.

- *Disinfectants*, which are essentially pesticides, are common in all-purpose cleaners. Skin contact and vapours can be irritating and corrosive to the skin and respiratory system. They are especially hazardous when dispersed from aerosol cans and sprays because the disinfectant can easily be inhaled via the nose and mouth. Common disinfectants include ammonia, bleach, pine oil, lye, cresol and phenol (see below). Such products are, at best, a temporary measure for making your home 'germ free'.

- *Ammonia* is a poison. Its fumes, even at the low levels contained in most glass cleaners, can irritate eyes, lungs and skin. Products with ammonia should not be used around children or elderly people with respiratory problems such as asthma since ammonia can aggravate these conditions. Splashes can cause burns or skin rashes. Never mix ammonia with chlorine-containing products since the mixture produces deadly chloramine gas.

- *Phenol*, also known as carbolic acid, causes central nervous system depression, can severely affect the circulatory system, is corrosive to the skin and is also a suspected carcinogen. Skin contact can cause ulceration, skin rashes, swelling, pimples, and hives. Phenol

---

Bleach Safety

Liquid household bleaches contain approximately 5 per cent sodium hypochlorite solution. Used properly it can be a simple and effective disinfectant. However, chlorine bleach fumes can also be highly irritating to the skin, eyes, nose, and throat. Contact with the skin can result in dermatitis. Ingestion can cause oesophageal injury, stomach irritation, and prolonged nausea and vomiting.

Bleach should never be mixed with any other cleaning solution. When mixed with acids such as ammonia, toilet bowl cleaners, drain cleaners or even vinegar it can release chloramine gas which can produce coughing, loss of voice, a feeling of burning and suffocation and even death.

---

is also an anaesthetic, so extensive damage to skin and tissues can occur before pain is perceived.

- *Pine oil* is derived from steam distillation of pine tree wood. It is a common ingredient in many household disinfectants and deodorants. In its concentrated form it is a skin irritant and may cause allergic reactions.

- *Colours and dyes* used in household cleaners can also be readily absorbed into the skin. The dyes used in these products are usually carcinogenic coal tar dyes. They can also contain impurities such as arsenic and lead which are also known to cause cancer.

- *Fragrance.* Because household cleaners contain strong detergents and other chemicals that do not smell particularly nice, manufacturers put very strong fragrances in them. These do not add anything to the cleaning power of a product but can irritate skin and lungs, and if not used cautiously can damage the very surfaces they are meant to clean.

Spray cleaners are basically the same mix of chemicals as liquid cleaners though often in a less concentrated form. Although pump sprays do not contain aerosols, when you use them you are producing a fine mist of chemicals that can easily be inhaled and quickly enter the bloodstream. So even though the product is less concentrated, you may inhale more chemicals. Amazingly, for a product that is mostly water, spray cleaners are also more expensive to purchase than standard liquid formulas.

Cream Cleaners

Scouring powders and abrasive creams are a traditional standby for shifting greasy deposits in the kitchen and soapy scum in the bath. These products are generally made with strong bleaches or ammonia to make them work faster. When the dried bleach mixes with water it produces chlorine fumes that are irritating to the eyes, nose, throat and lungs.

Powder Cleaners

Dry powders also contain *crystalline silica* which is an eye, skin and lung irritant and a carcinogenic when inhaled. It is all too easy to inhale silica particles when you sprinkle scouring powder in the sink or bath (and some of these types of cleaners are simply not suitable for certain surfaces like plastic because they can scratch). Once inhaled, mineral powders whether talc, silica, feldspar or quartz lodge more or less permanently in the lungs.

Alternatives

The best way to keep your home clean is to not let grease and grime build up in the first place. A quick regular wipe down is still the best way to prevent surfaces becoming sticky, greasy and grimy. It will also make big jobs like cleaning ovens and hobs a lot easier in the long run. However, not many of us can consistently reach this level of perfection.

To limit your exposure to the chemicals in household cleaners try the following options:

- *Use hot water.* Many people forget that hot water and steam are amongst the best, and most effective, household cleaners. In addition, it is usually the elbow grease, not the chemicals, that really gets the job done.

- *Dilute (and dilute again).* Most liquid detergents and soaps can be made into useful all-purpose cleaners simply by diluting them. To make sure you are being exposed to the minimum of chemicals try using an ecologically sound vegetable-based dish detergent or even better a liquid castile soap, well diluted in water.

- *Limit your use of sprays* (even pump sprays of mixtures you have made yourself) for hard-to-get-at places, to reduce the number of chemicals you inhale.

- *For tough grease* make a strong solution containing:

1/2 tsp of washing soda (also known as sodium carbonate, soda ash or sal soda)
2 tblsp distilled white vinegar
1/4 tsp liquid soap or dish detergent (or alternatively 1 tsp soap flakes)
2 cups of hot water

Add a few drops of essential oils if you wish. This can be used neat or put into a spray bottle to help you get at hard-to-reach places. Always wear protective gloves when working with washing soda.

- *Try making a simple scouring powder* from 200g bicarbonate of soda. That is all you need. Keep it in an airtight jar and use when you need to. If you like, add 10 drops of an essential oil and mix well to make a pleasant scent. Lemon, grapefruit, mandarin, tea tree, rose, peppermint or lavender are all suitable.

- *To make a stain-removing scouring powder* use:

8oz bicarbonate of soda
3oz borax
3 tblsp soap flakes (lightly crushed)

This will get rid of all but the most stubborn stains. To increase the bleaching power of this mixture try adding 3 tblsp of sodium perborate (available at healthfood shops). Sodium perborate is a bleach and you should wear gloves when using this formula, to avoid skin irritation.

- *An extra-strong cleaner* can be made with a mixture of liquid soap and trisodium phosphate (TSP). Add the following ingredients to 1 1/2 pints of warm to hot water:

1 tsp liquid soap
1 tsp trisodium phosphate (TSP)
1 tsp borax
1 tsp distilled white vinegar

TSP is a strong skin irritant so you should wear protective gloves when using it. The benefit of using TSP in this particular mixture, apart from its effectiveness against grease and mildew, is that you actually know what chemicals you are using.

Chapter 13

Bathroom Cleaners

Chemical Overkill

When you are cleaning your body, it is nice to know that you are doing it in a relatively clean space. What is more, in many busy households the bathroom is the only private place and becomes a temporary refuge. For these reasons many people pay special attention to cleaning the bathroom.

But, equally, because of the extra attention we pay to bathrooms, the smallest room probably has a greater concentration of toxic chemicals in the air and on its surfaces, than any other room in the house.

Many people think of the toilet bowl as the dirtiest part of the bathroom, but this is usually not so. Your toilet bowl has a constant flush of water running through it to keep it clean. The majority of really dirty jobs in the bathroom include cleaning areas such as the base of the toilet, underneath the seat, and under the rim (which can get pretty nasty). The soapy scum that builds up on the shower door is another chore and is usually the result of neglect. Some problems like mildew on the tiles are caused by poor ventilation and others, like the mineral deposits that clog showerheads and make faucets look unsightly are caused by living in a hard-water area.

Even so, with specialist cleaners you may be wasting your time and money, or simply be fooling yourself about how effective they are for cleaning your bathroom. You may also be exposing yourself unnecessarily to very harsh chemicals. Why risk it when nearly all bathroom cleaning jobs can be done with everyday cleaners and a few special mixtures of non-toxic ingredients?

ℒ

Toilet Cleaners and Disinfectants

We have become pretty obsessed with clean toilets in recent years. Spend a night watching television and you are likely to see commercials with animated toilet bowls singing about how happy they are now that they are using brand X, or cartoons showing flowers sprouting out of the bowl.

The reality is that liquid toilet cleaners contain many dangerous ingredients. As if to underscore the point it is a well-established fact that most household chemical accidents and poisonings occur in the bathroom.

The types of products available range from specialist cleaners (liquids and powders) to in-cistern devices that colour and fragrance the water as well as clean it, and the in-bowl devices that do the same. They are all made from roughly the same ingredients and the latter two types are perhaps the most insidious since every time you flush you produce a microscopic spray of easily inhaled chemicals into the air in your bathroom.

Toilet bowl cleaners are very aggressive and you have to wonder whether any toilet in any home really needs to be exposed to the amount of cleaning power typically provided by chemicals like:

> detergents/surfactants, strong acids such as muriatic (hydrochloric) acid or sulfuric acid (sodium bisulphate), oxalic acid, sodium hypochlorite, calcium hypochlorite, sodium carbonate, sodium metasilicate, phenols, ammonia, bleach, naphthalene, para-dichlorobenzene (PDCB), 5-dimethyldantoin, quaternay ammonium salts, colours and fragrance.

Toilet bowl cleaners usually contain strong cationic detergents which, in addition to cleaning power, also act as mild disinfectants. *Quaternay ammonium compounds* are typical of this type of surfactant. Cationic detergents can be toxic and poisonous to ingest;

they may cause nausea and in extreme cases coma. Cationics are also more easily absorbed by the skin, which is another reason why manufacturers recommend the use of protective gloves and advise thorough rinsing, should any of the product come into contact with the skin.

- *Sodium bisulphate* when mixed with water creates sulphuric acid. Sulfuric acid is highly toxic and corrosive to the skin and eyes. Its fumes are irritating to the airways, so adequate ventilation is vital when using toilet cleaners.

- *Oxalic acid* can damage the kidneys and liver. It is irritating to the eyes and respiratory tract and is corrosive to the mouth and stomach.

- *Muriatic acid*, also known as hydrochloric acid, is a severe eye, skin and mucous membrane irritant. It is highly toxic when inhaled and though lacking full safety data its effects are thought to be systemic.

- *5-dimethyldantoin* forms hypochlorite (bleach) when mixed with water. Hypochlorite is corrosive to skin and mucous membranes.

- *Para-dichlorobenzene (PDCB)*, in addition to being a common insecticide, is used as a mildew control agent. It is irritating to skin, eyes and throat and has been shown to cause liver damage in animals. Common symptoms of exposure to PDCB include drowsiness, weakness, nausea and headaches.

- *Naphthalene* is poisonous to humans and a potential carcinogen. It is a common pesticide, insecticide and fungicide and is also used in the manufacture of lacquers and varnishes. It is irritating to the skin, eyes, mucous membranes and the upper respiratory tract. Exposure can cause nausea, vomiting and profuse sweating.

- *Lye*, also known as sodium hydroxide or potassium hydroxide, is a highly caustic substance that burns the skin and can cause blindness. Products containing sodium hypochlorite (bleach) usually also contain lye to improve stain removal.

Toilet bowl cleaners with added disinfectants contain a number of additional hazardous ingredients, among them the following:

- *Ammonia*, even in low concentrations, can cause severe eye, lung, and skin irritation. Splashes can cause burns or skin rashes. Ammonia should never be mixed with chlorine-containing products since the mixture can produce highly toxic and sometimes fatal chloramine gas.

- *Bleach*. This can be an effective disinfectant. However, it can also be irritating to the skin, eyes, nose, and throat. Ingestion can cause oesophageal injury, stomach irritation, and prolonged nausea and vomiting.

- *Phenol*, also known as carbolic acid and cresol. It is a disinfectant that causes central nervous system depression, can severely affect the circulatory system, is corrosive to the skin and is also a suspected carcinogen. Light sensitivity and sinus congestion are common with exposure to fluids or vapours. Skin contact can cause ulceration, skin rashes, swelling, pimples, and hives. Phenol is also an anaesthetic, so extensive damage to skin and tissues can occur before pain is perceived.

- *Pine oil* is derived from steam distillation of wood from pine trees. It is a common substance in many household disinfectants and deodorants. In its concentrated form it is a skin irritant and may cause allergic reactions. If swallowed, pine oil may be sucked into the lungs (aspirated), possibly resulting in chemical pneumonia.

Alternatives

Not even the darkest recesses of your toilet need attacking with this many detergents and disinfectants. Once again it is not the cleaning products but the tools you use with them that are important. So the first thing to do is invest in a really good quality toilet bowl brush –

one with stiff bristles which can be stored in a unit that allows it to air dry in between uses. You can literally clean your toilet with any detergent and get the job done. Plain, diluted vegetable-based dishwashing liquid or castile soap is ideal. Or try making a more sophisticated mixture.

- Make a *simple toilet cleaner* with:

  4oz vegetable-based dishwashing liquid or castile soap
  2 cups baking soda
  2oz water
  2 tblsp white vinegar

You can even add ¹/₂ tsp essential oils of peppermint, lemon, pine, tea tree or eucalyptus to give it a fresh clean scent. Mix all the ingredients (adding the vinegar last) and put them into a *thoroughly rinsed* squeezable bottle. This mixture can be used inside and outside the toilet bowl.

- *To remove mineral deposits.* Add a cup (250ml) of white vinegar to the toilet bowl, then toss in a handful of baking soda. Let this bubble away for 10 to 15 minutes before giving the bowl a good scrub and flush.

- *An overnight soak.* A simple way to clean is to pour ¹/₂ cup of borax straight into the bowl, use your brush to give the bowl a quick once over, then leave it overnight.

- *Another overnight soak.* At night, place two 1,000mg Vitamin C tablets (unflavoured), or a mixture of 2 tblsp each of citric acid and bicarbonate of soda, in the toilet bowl. In the morning brush around the bowl and flush. This can help remove scum below the waterline.

If you still feel you must use conventional cleaners, make sure you put the lid down before you flush the toilet. That way you will not be sending a spray of chemicals into the air every time you flush.

Bath and Shower Cleaners

These are usually nothing more than all-purpose cleaners with a few extra ingredients, usually acids. Typically they come in liquid, powder and cream form. Their job is to cut through the soapy scum that modern detergents (so we are told) are not supposed to leave behind on tubs, tiles and glass shower doors.

To make safe the use of conventional bath and shower cleaners, never use aerosols, always wear gloves, and make sure the room is well ventilated. Never use these products around food, children or animals. Once again, unless your bathroom is sorely neglected you really do not need special cleaners to get the job done.

- *Use natural disinfectants.* Both white vinegar and lemon juice act as mild disinfectants. Because they are acidic they can be used to remove hard-water spots, dissolve mineral build-up and break down filmy soap residue without leaving a film of their own. They also act as deodorisers. Use two parts water to 1 part vinegar or lemon juice or use undiluted for heavily soiled areas. Since both vinegar and lemon juice are acids remember to wear rubber gloves if you are going to be using either for more than a few minutes. For scrubbing down the bathroom a nylon scrubbing pad is a must.

- *A simple abrasive.* Baking soda can be used neat to provide a mild abrasive action that is safe for tiles, counter tops, sinks and tubs. Use it with a damp cloth on surfaces or use it with an old toothbrush to reach hard-to-get-at places.

Buy shower curtains that are washable, and wash them regularly to prevent soap and mildew build-up. If they get really scummy, spraying with undiluted distilled vinegar before washing will help remove soap residues. If you have a glass shower door wipe it down after each use with a super absorbent sponge cloth, or better yet with a squeegee which will be even more efficient. Wiping down tiles and glass doors regularly will prevent soap scum from building up. Sounds like too much work? Actually it only takes about 30 seconds.

Limescale Removers

Bathroom descalers are similar to those used in kettles and are usually based on either citric acid or phosphoric acid. Citric acid is relatively mild and harmless while phosphoric acid can cause skin and eye injuries.

For a less toxic way to remove limescale in the bathroom consider the following:

- *Metal showerheads* can be cleaned by boiling them for 15 minutes (or until the limescale deposits begin to soften) in half a cup (125ml) of white vinegar and one litre of water. This will remove all but the most stubborn mineral deposits. Scrub clean with an old toothbrush.

- If you have a *plastic showerhead* soak it in equal parts white vinegar and warm water for one hour, then scrub clean with an old toothbrush.

- To remove mineral deposits around *faucets*, lay a cloth soaked in hot vinegar on the fixture that needs cleaning. Let it stand for an hour or so, and wipe or gently scrub off.

Mildew Cleaners

Mildew removers come in many forms; in liquids that you rub on and rinse off, and more recently in sprays that you do not rinse off. They all contain the same basic ingredients. Among these are *phenol* – a central nervous system depressor, skin irritant, circulatory system disrupter and suspected carcinogen (see above for more); *kerosene* – which can dry the skin and cause lung inflammation, and *pentachlorophenol* – which is toxic when inhaled, absorbed or ingested and is a known reproductive toxin associated with birth defects.

Most mildew cleaners only work in the short term. So if you are going to use a short-term solution, why not make it a less toxic one?

- *Borax* diluted in a little hot water and applied with an old toothbrush will remove mildew just as effectively. If the problem is really severe, leave the solution for an hour or so before rinsing off.

- *Think airflow.* Mildew is most likely to build up in damp places so if you are having a problem, poor ventilation may be the cause.

Chapter 14

Carpet and Upholstery Cleaners

Tread carefully

Every time you tread on your carpet, you send an invisible cloud of dust, dirt and chemicals into the air. Most people do not realise how effective their carpets are as dust and pesticide traps. But there are studies to show that, in more than 75 per cent of the homes tested, rug dust contained high levels of mutagenic levels (substances that alter cellular DNA; mutagens are often precursors of cancer) and toxic levels of chemicals. Infants and toddlers ingest 2.5 times more of these than adults do because they spend so much time sitting and playing on the floor. For this reason keeping your carpet clean is probably more important than you know.

Whether carpet shampoos are the best way to keep carpets clean, however, is debatable. Carpet and rug shampoos are concentrated detergents that come in ready-to-use liquids, trigger sprays, powders and aerosols. They are formulated to wet the pile of the carpet and take up superficial oil and grease. Applied to the carpet they attract and trap soils and then dry to a brittle powdery residue that can be removed by vacuum cleaner.

All commercial carpet cleaning solutions can leave chemical residues and odours. Whether you are using a spot clean spray or a detergent in a carpet-shampooing machine, the basic ingredients are much the same.

Carpet shampoos typically contain water and a mixture of other ingredients like detergents and surfactants, petroleum-based solvents such as butyl cellosolve and preservatives. Some contain enzymes to remove protein stains, optical brighteners, soil-retardants such as formaldehyde, perfumes and deodorisers.

- *Detergents and surfactants* used in products such as carpet shampoo can be contaminated with 1,4-dioxane, a carcinogen. As labels do not always disclose the type of detergent used, it is best to assume they are toxic until proven otherwise.

- *Solvents* are neurotoxic and associated with birth defects and miscarriages. They are quickly absorbed into the skin, which is why you should always wear rubber gloves when using solvent-based cleaning fluids. Inhalation can cause headaches and nausea and damage to internal tissues and organs such as the liver and kidneys.

- *Formaldehyde* is a suspected human carcinogen that (among other things) alters your sense of smell and can cause respiratory irritation. Anyone with asthma, lung infections or similar ailments can be severely affected by this preservative. It can also cause stuffy nose and red, itchy, teary eyes, nausea, headache, and fatigue. Some of the other preservatives used in household detergents, such as quaternium compounds, can also break down into formaldehyde.

- *Optical brighteners* do not clean the carpet. Instead they remain on top to create an optical illusion. Like thousands of minuscule mirrors they reflect light back to the eye, giving the impression that the carpet is cleaner and brighter than it is. Optical brighteners can cause skin sensitisation and allergic reactions.

- *Fragrances* used in carpet cleaners are very strong and are included to cover up unwanted smells. Synthetic fragrances can linger and be highly irritating to the respiratory tract and may cause symptoms such as headache, nausea, short-term memory loss and disorientation and an inability to concentrate.

Upholstery cleaners are similar in their formulation to carpet cleaners and can produce the same toxic effects. All commercial carpet and upholstery cleaning solutions leave chemical residues on your carpet and soft furnishings. This residue is perhaps the greatest risk to young children who tend to (a) crawl around on carpets and soft furnishings and (b) often stick their hands in their mouths.

> Carpets become really dirty. Actually they come into your home pretty 'dirty'. The average modern carpet contains over 200 different noxious chemicals including flame retardant and stain repellants. It can also off-gas formaldehyde and other toxic gasses for up to 20 years. In one study that examined the neurotoxic effects of carpet off-gassing on animals, 90 per cent of carpet samples tested produced at least one toxic effect and 60 per cent produced 3 or more 'severe neurotoxic effects'.

Spot Removers

Products intended to spot-clean carpets are at best a short-term solution for stains and at worst a toxic hazard. Most of these are poisonous when ingested, highly irritating when inhaled, and can cause skin reactions in sensitive people. Typically they contain the following:

- *Perchloroethylene*, a solvent and carcinogen. It is a common constituent of dry cleaning fluids and has been shown to cause liver, kidney and central nervous system damage.

- *Trichloroethylene* is another commonly used solvent and a suspected carcinogen. It is associated with eye, skin and lung irritation, central nervous system disorders and damage to vital organs such as the kidneys and liver. It can also cause cardiac failure. Trichloroethylene can persist on fabrics long after its application. It is also a common groundwater contaminant that easily works its way back into the food chain.

- *Diethylene glycol*, a solvent belonging to a family of chemicals called *dioxanes* (not to be confused with the pesticide dioxin). Dioxanes are thought to be carcinogenic and some research suggests that they are immune system suppressants.

- *Ammonium hydroxide*, a relative of lye; is corrosive and very irritating to eyes, skin and respiratory tract.

- *Naphthalene*, also known as tar camphor. It is poisonous to humans and is a major ingredient of pesticides, insecticides and fungicides. It is also used in the manufacture of lacquers and varnishes. It is irritating to the skin, eyes, mucous membranes and the upper respiratory tract. Exposure can cause nausea, vomiting and profuse sweating. Prolonged exposure can cause damage to the kidneys, liver and red blood cells, and possibly cancer.

- *Oxalic acid,* a poisonous organic compound commonly used in household cleaners. It is also commonly used in cleaning automobile radiators and other metals. It can damage the kidneys and liver, irritate eyes and respiratory tract, and is corrosive to the mouth and stomach.

- *Perfumes* derived from synthetic sources. Such perfumes may be combinations of hundreds of unknown chemicals. Many synthetic perfumes are associated with central nervous system problems, with allergic and asthmatic reactions, headaches, nausea and lethargy.

Alternatives

Regular steam cleaning is the best way to keep your carpet toxin-free. It may be expensive but some may reason it is worth the money since, as well as cleaning, the very hot water used in steam cleaners kills dust mites, fleas, bacteria and mould.

Throw-rugs should be laundered regularly, if possible. Non-washable rugs should be shaken outside or hung on a line and beaten with an old-fashioned rug beater or old tennis racket.

In between steam cleans you can help keep your fitted carpets clean by:

- *Removing shoes* before you come into the house. Most dirt is tracked in on the bottom of shoes.

- *Vacuuming regularly.* If possible aim to vacuum weekly to keep the amount of dust and dirt you send into the air to a minimum.

- *Using a carpet sweeper* in between vacuums. These are inexpensive and great for picking up little spills and bits of unidentifiable stuff that falls to the floor.

- *Spot-cleaning* carpets and upholstery with a solution of $1/2$ cup borax in one litre of warm water. Use a stiff bristle brush for carpets and a softer one (like a toothbrush) for upholstery stains. Vacuum well when dry.

- *Buying washable soft coverings* (cushion covers, etc.) and laundering them regularly. If your sofas and chairs get a lot of wear, consider a washable throw for everyday protection.

Carpet Deodorisers and Fabric Fresheners

Carpet fresheners or deodorants are mostly heavy perfumes formulated to cover up odours in floor coverings. Most commercial carpet deodorising powders are not at all sophisticated. Usually, they are made by using a baking soda base and a strong petrochemical-derived fragrance.

Fabric fresheners are similar to carpet deodorisers. These products are often marketed to smokers and pet owners as a way of removing nasty smells from their soft furnishings. They are mostly made up of surfactant, which allows the chemicals to penetrate deeply into the fabric, and perfume. They do not remove smells. Instead, like air fresheners, they contain chemicals that alter your sense of smell. It is anyone's guess what is in the thin coating of surfactant that builds up on your soft furnishings.

Given that some commonly-used surfactants are contaminated with carcinogens like 1,4-dioxane, using them hardly seems worth the risk. Prevention is still the best way to avoid carpet and upholstery smells. Or alternatively:

- *Make your own*. You can make your own carpet deodorant by using plain baking soda fragranced with essential oils. Sprinkle this on your carpet and work well into the pile. Leave anywhere from 15 minutes to overnight depending on the strength of the odour, and then vacuum. This is less expensive than buying commercial brands and just as effective.

- *Disinfect* carpets with borax. Sprinkle on your carpet, work into the pile and wait anywhere from two to 24 hours before vacuuming to remove the powder. Really mouldy or foul-smelling carpets may need more than one application.

  Note: Inhaling any powder can cause respiratory problems in some individuals, so make sure you use other rooms while these minerals are soaking in.

- *Open a window*. It is still the best way to deodorise most things in your home.

Chapter 15

Dish Detergents

Clean Enough to Eat Off?

While not on the same level as laundry detergents, washing-up liquids can still add to your total toxic load, especially if you are using hot water and no protective gloves. Because dishwashing detergents are generally made from anionic and non-ionic surfactants they are less well absorbed than some other types of household cleaners and so are considered to be low in toxicity. Nevertheless, anionic and non-ionic detergents can still cause irritation to the skin, eyes and mucous membranes.

> Most dishwashing liquids are a mixture of water, detergent, surfactant, preservatives, alcohols, petroleum distillates, ammonia, salts, colouring and perfume.

- *Detergents and surfactants* such as diethanolamine (DEA) and triethanolamine (TEA) and morpholine may mix with formaldehyde forming preservatives to form nitrosamines. Other types may be contaminated with 1,4-dioxane. There are other problems with detergents as well.

  Some detergents make use of a class of surfactant known as alkyl compounds. Alkyls are problematic in many ways. One common type, the alkyl benzyl sulfonates (ABS) group, includes linear alkyl benzene sulfonates (LAS) and linear alkyl sodium sulfonates that are slow to biodegrade. LAS are the most common surfactants in use in household products.

  Another type, alkyl phenols (which include nonyl phenol, nonyl phenoxy ethoxylate, alkyl phenol polyglycol and poylethylene glycol alkyl aryl ethers) are hormone disrupters. One member of this family of chemicals is the spermicide nonoxynyl (used on condoms and in spermicidal foams and gels) which may give you some idea of their powerful biological toxicity.

The environmental impact of these substances has already been noted in scientific studies. Many animals have experienced reproductive problems because of them. Humans are also likely to be affected by alkyl phenoxy compounds and it is thought that they are particularly harmful to pregnant women and young children. Prolonged exposure may cause endocrine disruption that can affect reproduction, menstrual cycles, thyroid, pancreas and other hormone-producing organs and may even contribute to estrogen-dependent cancers.

Alkyl phenols are still used in commercial products in the US, but are confined to industrial detergents in the UK. They can get into the water supply and food chain and are thought to contribute to reproductive problems in men such as low sperm counts, damaged sperm and testicular cancer.

- *Petroleum distillates.* The chemicals used in most washing-up liquids to make the surfactants more water-soluble are known as ethoxylates. These are based on ethylene oxide – a petroleum derived product. Ethoxylated alcohols, which are just as prominent in natural as in conventional brands, can be contaminated with the carcinogen 1,4-dioxane. They may also contain diethanolamine (DEA) which is a nitrosamine precursor. Some petroleum distillates such as PEG are used as preservatives. They can be formaldehyde formers that can mix with DEA and TEA to form carcinogenic nitrosamines.

- *Ammonia* vapours, even in low concentrations, can cause severe eye, lung, and skin irritation. Splashes can cause burns or skin rashes. Ammonia should never be mixed with chlorine-containing products since the mixture can produce highly toxic, even fatal, chloramine gas. Quaternay ammonium compounds are part of the ammonia family. They are formaldehyde-forming preservatives that can produce nitrosamines when mixed with DEA and TEA.

- *Colours* used in dishwashing liquids are usually coal tar colours that are carcinogenic. Other colours can be contaminated with carcinogenic substances such as arsenic and lead. Dye can easily penetrate the skin, especially if it has been soaking in hot water. Even though you cannot see it, colours can also leave residues on your dishes and utensils. As dyes do not add to the cleaning power of a product you might wish to avoid coloured dishwashing products.

- *Fragrance.* Because household cleaners contain strong detergents and other foul-smelling chemicals, formulators put very strong perfumes in them. Fragrances of course do not add anything to the cleaning power of a product and this heavy concentration of chemicals can irritate skin and lungs.

Alternatives

There are probably more urgent things to worry about than your dishwashing liquid and in truth few good alternatives exist.

To make using dishwashing detergents safer for the environment, choose products that exclusively use vegetable-based detergents, but do not be fooled that these are always free from harmful chemicals. In addition:

- *Use less.* Adding a little borax to the water beforehand will make even the smallest amount of washing-up liquid more effective. This is because borax is an effective water softener.

- *Wear rubber gloves* to prevent irritation of the skin and to prevent the absorption of contaminants in the water including nitrosamines, dioxanes, dyes, colours, perfumes.

- Some washing-up liquids contain small amounts of *ammonia*, so do not mix washing-up liquids with bleach or bleach-containing products as this mixture will form a deadly chloramine gas.

- Avoid products that are *coloured* or *fragranced*.

- *Dilute your regular brand*. Most dishwashing liquids are very concentrated and we all tend to use much too much to get the job done.

Diluting will not substantially affect the cleaning power of the product. So try mixing equal amounts of vegetable-based washing-up liquid and water. The mixture can be thickened with approximately 1 tablespoon of salt. Store in a squirt-top bottle (any well-rinsed dishwashing detergent bottle will do). This will still foam and will still clean your dishes but will expose you to fewer chemicals.

To make an even less toxic version of this recipe replace conventional dish detergent with castile soap. Dr Bronner's scent-free liquid baby soap is a good choice. That way you can add your own fragrance with essential oils if you wish.

As an alternative to using dish detergent, wash your dishes in *soap flakes* with a dash of vinegar to cut through the grease. You will not get loads of bubbles but your dishes will be clean. You can also wash your dishes effectively in *washing soda* or *borax* (always wear gloves when using these substances).

- *Rinse your dishes well* before leaving them to dry. Soaping them up and leaving them to drain means that chemical residues get into the next serving of food you eat from that dish.

Automatic Dishwashing Detergents

Most automatic dishwashing detergents are either irritants or corrosives depending on their composition, concentration and physical form. They are made with strong detergents and a strong alkali, and so have a pH value of between 10.5 and 12.0. Skin irritation or burns may occur following exposure to dissolved detergents. Many also contain washing soda (sodium carbonate) and occasionally phosphates that pollute waterways. Additionally, they may contain sodium silicate to prevent damage to the dishwasher, a surfactant to prevent spotting, and fragrance.

Some automatic dishwashing detergents also contain dry chlorine that becomes activated when mixed with the water in the dishwasher. This means that when you open the dishwasher, chlorine fumes are released in the steam that leaks out. These can cause eye irritation and difficulty in breathing, especially for those with respiratory problems.

The average individual is unlikely to have prolonged contact with these substances. However, if you have children in the house you might consider using a powdered variety rather than the tablet or liquid variety. Tablets have become a significant source of poisoning among children and animals so if you choose these, store them carefully.

Alternatives
All automatic dishwashing detergents will leave some chemical residues on your dishes. However, you can cut your exposure to these by trying the following:

- *Choose powdered brands*, which, generally speaking, contain fewer harmful ingredients than the liquid variety. Powders are also less likely to be mistakenly swallowed by young children.

- *Use less.* The powder receptacles in dishwashers are larger than they need to be and encourage us to put more detergent than necessary into the machine. The average dishwasher only needs about 2 tblsp of detergent in the closed cup dispenser, and 1 tblsp in the open dispenser, to work well.

- *Use low or no phosphate brands.* These are much kinder to the environment.

- *If commercial products are unacceptable* try using 1/2 teaspoon of pure soap powder (more will produce too many suds), mixed with 3 tablespoons of baking soda.

- *For normally soiled dishes* use 2 tsp borax pre-dissolved in hot water. Then use vinegar in the rinse cycle.

- *For heavily soiled dishes* try washing with 1/4 cup of sodium hexametaphosphate – a relative of trisodiumphosphate (TSP).

- *Sprinkle your dishes with baking soda* before loading them in the dishwasher. Then add detergent to the closed dispenser only. This will cut down on detergent use and you will still get clean dishes (Note: do not put baking soda on your aluminium pans – it will discolour them).

Chapter 16

Floor Cleaners

Toxins Underfoot

Floor cleaners are generally made from either petroleum distillates and solvents or water-based detergents. The mixture makes a cocktail that is dangerous to inhale and can irritate the skin. Of the two types, water-based cleaners are much less toxic. Those containing the solvent glycol monomethyl ether are much more hazardous since, when swallowed, they can cause nausea and vomiting, stomach pain, bleeding, and/or chemical pneumonia.

Those that utilise pine oils can cause convulsions, coma and even death if ingested. While this is unlikely because of their strong taste, parents of very young children might reason that, given good alternatives, keeping such products in the house is just not worth the risk.

Of course, floor cleaners do not have to be swallowed to be toxic. If you have young children who tend to crawl around on the floor, they may be picking up these substances on their hands and breathing in their fumes. Prolonged and repeated skin exposure and exposure to the vapours from floor cleaners can result in central nervous system depression, producing symptoms similar to drunkenness and kidney injury.

> The average bottle of floor cleaner contains detergents, either ammonia or bleach, phosphates, pine or citrus oils. Others contain synthetic solvents such as butyl cellosolve, phosphoric acid, potassium hydroxide, colour and perfume. Most contain undisclosed preservatives as well.

- *Detergents.* As in all detergent-based products there is a risk that the detergent is either contaminated with the carcinogen 1,4-dioxane or is a planet-poisoning alkyl compound.

- *Solvents* can be flammable, toxic, or pose a serious health risk through skin absorption and inhalation. Some health hazards occur immediately. Others, such as liver and kidney problems, birth defects, and nervous disorders, occur slowly over time. In addition, many solvents adversely affect the central nervous system, producing drunken or narcotic effects that can permanently affect normal bodily functions.

- *Bleach* can be a good disinfectant. However, bleach liquid and vapours are irritating to the skin, eyes, nose, and throat. Skin splashes can cause dermatitis and ingestion can cause oesophageal injury, stomach irritation, and prolonged nausea and vomiting.

- *Ammonia* vapours, even in low concentrations, are severely irritating to the eyes, lungs, and skin. Skin contact can produce burning and rashes. Ammonia should never be mixed with chlorine-containing products since the mixture can produce highly toxic chloramine gas that can cause coughing, loss of voice, feelings of burning and suffocation, and even death.

- *Phosphates* are used to soften the water and help the detergents rinse better. However, they are also devastating for the environment, causing the algae blooms that choke rivers and other waterways.

- *Potassium hydroxide*, a form of lye, is a caustic product that burns the skin and can cause blindness.

- *Pine oil* is derived from steam distillation of pine tree wood. It is a common ingredient in many household disinfectants and deodorants. In its concentrated form it is a skin irritant and may cause allergic reactions.

- *Colours and dyes* can also be readily absorbed into the skin. The dyes used in these products are usually coal tar dyes that are carcinogenic. They can contain impurities such as arsenic and lead, also known to cause cancer.

- *Fragrance.* Very strong fragrances are used to cover up the odour of detergents and other chemicals that do not smell particularly nice. Fragrances, of course, do not add anything to the cleaning power of a product, but can irritate skin and lungs and cause central nervous system disorders.

Floor Polishes

Floor polishes add a temporary shine to your floors but at a price. Commonly they contain plastics that can damage vinyl tiles and turn them yellow over time. In addition, they contain solvents, petroleum distillates, naphtha, nitrobenzene and ammonia.

- *Petroleum distillates* refer to a broad range of compounds that are a by-product of distillation during the refining of crude oil. Typically this family includes solvents (see naphthas below), surfactants, mineral oils and waxes. Use products containing petroleum distillates carefully. Do not mix them with other products, even other petroleum distillates. Always wear gloves and avoid breathing vapours (open a window if you can). Keep petroleum distillate products out of reach of children.

- *Naphthas* are solvents derived from both petroleum distillation and coal tar. They are flammable, skin irritants, and toxic. Naphthas are often used as a base for insecticide. They can enter your system through inhalation and vapours, ingestion, and eye and skin contact. Repeated contact can cause skin irritation. Over-exposure may cause sensitivity to light and central nervous system depression. When using a product that contains naphtha, always wear gloves and open windows.

- *Nitrobenzene,* a highly toxic solvent, can cause spleen and liver damage. It can enter your body through inhalation, skin and eye contact, and ingestion. It affects the central nervous system, producing headache, fatigue, general weakness, vertigo and in some cases severe depression, unconsciousness, and coma. Drinking alcohol increases the toxic effects of nitrobenzene.

Nitrobenzene quickly crosses the placenta and can cause birth defects. If possible, avoid products that contain nitrobenzene.

How can a floor legitimately be called 'clean' when it has residues of all this hazardous waste on it? The silly thing is that you do not need anything special to clean your floors. The average household floor attracts average household grime and dust. Cleaning this off is a fairly simple task that can be made much easier if you make a habit of wiping up spills as they occur.

Alternatives
As with all toxic cleaners the best alternative is always to use less. Most of these products are highly concentrated and we all tend to overestimate how much we need. To make the use of conventional cleaners safer always wear gloves and always open a window. Or consider these alternatives:

- *A simple floor cleaner* can be made from hot water and vinegar. Add one cup of vinegar to a pail full of water. If your floor is really dirty mix a small amount of liquid soap or soap flakes with the hot water. Rinse afterwards with a vinegar and water solution. This works well on ceramic, polyurethane-finished wood floors, linoleum and other vinyl-tiles.

- *To disinfect*, mop the floor with $1/2$ cup of borax in a pail of hot water. To make sure the borax dissolves thoroughly, put the borax in first and then add the water, stirring until it is dissolved.

- *To cut grease*, mix $1/2$ cup washing soda (sodium carbonate or sal soda) with 1tblsp liquid soap, $1/4$ cup distilled white vinegar and $1/2$ bucket of water. Washing soda is caustic so use gloves and avoid splashing. Do not use on waxed floors as it will mottle the finish.

- *For wooden floors*, use $1/4$ cup liquid soap with $1/2$ to 1 cup distilled vinegar or lemon juice to a bucket of water. To improve the fragrance use a cup of strong, freshly brewed herb tea of your choice.

- *For linoleum* combine 6 tblsp of cornstarch for every cup of water, mix in a bucket and use as you would any floor polish.

Chapter 17

Glass and Window Cleaners

Sparkle and Swoon

The solutions that we use to clean our windows and mirrors are unbelievably strong given that they have a simple job to do. Once again this is a result of manufacturers responding to what they perceive as consumer demand for fast-acting products.

Apart from paying a premium price for something that is mostly water, you are also exposing yourself to a number of chemical hazards. Window cleaners are inevitably sprays and because you work so close to the glass or mirror while cleaning you cannot help but inhale the mixture. Not surprisingly, glass and window cleaners can be mildly irritating to the eyes, skin, nose and throat.

Typically a glass and window cleaner will contain a high proportion of water and a mixture of other ingredients including: solvents such as isopropyl alcohol or butyl cellosolve, ammonia, glycol ether, silicon, waxes, formaldehyde and, if it is an aerosol, propellants such as isobutane or propane and sometimes colours and perfumes.

- *Solvents* have been associated with liver and kidney problems, birth defects and nervous system disorders. Many of these problems develop slowly over time and are thought to be the result of chronic low-level exposure to solvents such as butyl cellosolve, ethyl or isopropyl alcohol, propylene glycol and ethanol. The solvents used in household products are referred to as 'organic' solvents – not because they are free from chemicals but because they contain the carbon atoms common to all life on earth. All organic solvents are hazardous and are easily absorbed via the skin and via their fumes.

- *Ammonia* is a poison. Its fumes, even at the low levels contained in most glass cleaners, can irritate eyes and lungs. Products with ammonia should not be used around children or elderly people with respiratory problems such as asthma since ammonia can aggravate these conditions. Splashes can cause burns or skin rashes. Never mix ammonia with chlorine-containing products since the mixture produces deadly chloramine gas.

- *Propellants* are found in all aerosol sprays. They create fine particles of spray that are easily and deeply inhaled into the lungs and quickly absorbed into the bloodstream. Chemicals that may be relatively 'harmless' on your skin (because they are poorly absorbed) may become extremely dangerous if inhaled as a mist. Common propellants include HCFCs (hydrochlorofluoro-carbons), propane, butane and isobutane. Propellants are associated with nervous system disorders, skin, eye, throat and lung irritation, lung inflammation and liver damage. They are also highly flammable.

- *Colouring* is added to appeal to your senses. It does not add to the cleaning power of the product. Don't believe it? You are not alone. Consumer surveys have shown that many people actually believe that blue window-cleaning solutions clean better than clear ones.

Alternatives

If you are stuck on using conventional products, always use them in a well-ventilated room, and choose pump sprays over aerosols.

For most glass and mirrors in the home, water is the best cleaner. Consumer reports show that water cleans mirrors and windows better than some 60 per cent of the products on the market. Equally most homemade mixtures can equal and in some cases surpass commercially available products. Additionally, they do it at a fraction of the price of commercial brands.

Whatever you end up using to clean windows and mirrors will be made more effective and easier with the right tools. Make sure you have two lint-free cloths to do the job – one for cleaning and one for drying.

Most people just use paper towels, which are equally effective, but they are an expensive and wasteful way to clean. Some of the best and least expensive cleaning cloths are the big muslin squares that are commonly used as nappy liners. You can buy these in practically any mother and baby shop. Or if you want a bit of luxury, invest in the pure linen scrims used by professional window cleaners.

- *Simple glass cleaner 1.* Put some plain club soda in a spray bottle. Many people swear by this cleaner which appears to work because it contains sodium citrate; this acts like a water softener and helps the water to clean more effectively. It will not dry as quickly as conventional cleaners, but for lightly soiled glass and mirrors this is a perfectly adequate cleaner.

- *Simple glass cleaner 2.* Mix equal amounts of water and distilled vinegar in a spray bottle. The vinegar will help cut through greasy fingerprints and will also act as a room deodoriser.

- *Foaming glass cleaner.* If you like some sort of foam to show you where you have sprayed, or your windows or mirrors are particularly greasy and dirty, you can make up a stronger mixture with:

  $1/2$ tsp liquid soap
  3 tblsp distilled white vinegar
  2 cups of water

  Put the ingredients into a spray bottle and shake well. This works well inside the house (and is also an effective all-purpose cleaner for light cleaning) but can sometimes cause streaks on outside windows.

Windshield Wiper Solution

Windshield wiper solution is more hazardous than most of us realise. Most types contain anything from 37–100 per cent solvent such as ethylene glycol, isopropanol and methanol, the rest being made up with detergent and water. The most toxic windshield wiper solutions contain 100 per cent methanol.

- *Methanol* damages the nervous system, liver, kidneys; inhalation can lead to lung disease, and ingestion can cause blindness.

- *Ethylene glycol* poisons animals that are attracted to the sweet smell; can cause damage to internal organs through skin absorption; inhalation can cause dizziness.

- *Isopropanol* irritates mucous membranes; ingestion results in drowsiness, unconsciousness and death.

Due to its hazardous nature, windshield wiper solutions should always have a child-proof safety cap. Needless to say they should be stored away from children and pets. There are few real alternatives to their use, however, and when you are streaking along on the highway on a murky, muddy, rainy day, the safety of a clean windscreen is certainly a top priority. So, instead, try to choose solutions that are detergent-based. When topping up your car wear gloves and try to avoid skin contact and inhalation. Make sure the safety cap is replaced securely.

Chapter 18

Laundry Products and Fabric Care

What's in the Wash?

Wearing clean clothes and sleeping on clean sheets is a pleasure. And in the supermarket today you can find just about everything you need to make sure that your clothes are clean and bright. But as with all cleaning products there is a downside to having whites that are whiter than white.

Detergents were first developed as a result of the petroleum industry trying to find a way to make money from the toxic waste materials that they generated each year. One of these was propylene. Eventually, scientists found a way of mixing propylene with benzene to produce sulphuric acid. They then added sodium hydroxide to the mix to neutralise this harmful acid. A sodium salt, somewhat like ordinary soap, was the result and hey presto! the detergent industry was born. Today's detergents are still made largely from the waste materials generated by the petroleum industry.

- *Detergents/surfactants.* Detergents are responsible for making foam and lifting dirt from clothing. Surfactants are responsible for helping the other ingredients, as well as the wash water, penetrate more deeply into the fabric. Both can be contaminated with impurities that are carcinogenic, or can contain the

> Laundry detergents are a complex mixture of substances including detergents and surfactants, enzymes, bleaches, optical brighteners, builders/water softeners (pH adjusters) and processing aids, corrosion inhibitors, anti-redisposition agents and fragrances. Powdered varieties also contain fillers – inert substances that keep the powder flowing freely. Some powders are as much as 50 per cent filler.

carcinogen 1,4 dioxane. Alkyl compounds, which are commonly used detergents, are known pollutants. These chemicals are unlikely to be absorbed directly into your skin from washed clothes. Instead they threaten health by polluting groundwater supplies and entering the food chain.

- *Bleach* such as sodium perborate and sodium percarbonate can be an effective way to make whites look white. However, it is often used in such small amounts in laundry powders and liquids that it is not very effective (see Laundry Boosters below). Interestingly, bleaches do not remove stains at all. Instead they convert them into an invisible form through a process called oxidation. Bleach can react with other chemicals in the mix. It can also let off vapours that can irritate the lungs. Bleach residues on your clothes can cause skin reactions as well as damaging cloth fibres.

- *Optical brighteners* belong to a family of compounds that were once used in food products like flour and sugar to make them look whiter. They have long since been banned for this purpose. Optical brighteners put a thin chemical film on your clothes that converts ultraviolet light to visible blue light, making clothes look brighter and making the bleach appear to work better than it has. Optical brighteners can cause skin sensitisation and allergic reactions.

- *Phosphates* are commonly used as builders/water softeners. They dissolve hard-water minerals and help to increase the effectiveness of the detergent. They also act as anti-redisposition agents (see below). Phosphates, while relatively non-irritating, are an ecological disaster responsible for the overgrowth of marine plants that choke waterways and kill other forms of marine life. While manufacturers argue that phosphates make detergents more economical to use, detergents with phosphates really should be avoided. Other types of builders include polyphosphates, citrates, sodium carbonate (washing soda) sodium silicate, aluminosilicate (zeolite).

- *EDTA*, or ethylene diamino tetra acetic acid, is sometimes used as an alternative to phosphates. EDTA is a preservative that stabilises the bleach and foaming agents in detergent products, preventing them from becoming active before they are immersed in water. It is also used as a water softener. Unfortunately, it binds with toxic metals in the environment and remobilises them, carrying them back into our drinking water supplies and food, especially fish and shellfish. EDTA can be irritating to skin and mucous membranes and may cause allergic reactions. Liquid detergents are more prone to contamination and therefore use more preservatives than the powdered variety.

- *Anti-redisposition agents* stop dirt from getting back into your clothes. Commonly used are sodium carboxymethylcellulose, polyethylene glycol (PEG) and polyacrylates. PEGs can be contaminated with the carcinogen 1,4-dioxane.

- *Perfumes* do not do anything but smell nice. The fragrance portion of laundry powder is formulated to stay in your clothes. It can be a cause of skin and respiratory irritation and may cause headaches and other neurological symptoms in some individuals.

Enzymes

Most laundry detergents are formulated with enzymes to remove the most common stains. These products are called 'biological' detergents. Enzymes are obtained from selected strains of bacteria. Products that contain enzymes can be irritating to the skin and cause allergic reactions such as asthma and dermatitis.

Although the addition of enzymes is promoted as the best way to remove 'protein' stains such as grass, chocolate and blood, this is somewhat misleading. The most common clothing stains are not protein stains but dirt and grease stains. When you use these specially formulated products you are often paying for the money they have spent advertising unnecessary additives on TV.

Enzymes don't have to touch your skin to cause an allergic reaction. In December 2000, *The Lancet* reported an outbreak of asthma in a detergent factory due to inhaling airborne enzymes.

- *Corrosion inhibitors* protect the washing machine. Sodium silicate is a commonly used corrosion inhibitor.

- *Processing aids* help the main ingredients work more efficiently together. They can be builders such as phosphates and EDTA (see above). They can also be neurotoxicsolvents like isopropyl alcohol as well as xylene sulphonate and sodium sulphate.

Laundry detergents can also sometimes contain disinfectants such as pine oil that is highly irritating and may cause allergic reactions, and phenolics and coal tar derivatives that are carcinogenic.

Liquid detergents may contain extra ingredients such as opacifiers to make them look nice. These give the liquid a rich, creamy, opaque appearance but do not clean your clothes They also contain synthetic colours. Liquid detergents contain no filler at all and are more concentrated than powders. Other than that there is no difference between types of laundry detergents. Liquids do not clean better, they do not get into the wash quicker or penetrate fabrics more efficiently. They are just a variation on a theme.

Fabric Softeners

Fabric softeners are made from mild detergents and cationic surfactants such as quaternay ammonium compounds. They have a strong positive charge designed to stick to negatively charged wet fabrics. These surfactants form a uniform, positively charged layer on the surface, making it feel softer to the touch (excess negative charge on the fabrics is responsible for a scratchy feeling after washing and for static cling). Cationic surfactants are also used in laundry products that boast fabric-softening properties.

Fabric softening sheets release a special resin in the dryer that deposits a waxy coating on the clothes to make then feel softer.

Along with surfactants and resins, fabric softeners deposit a range of other chemicals on to your clothes as well. According to a report by the US Environmental Protection Agency, fabric softeners and dryer sheets contain an enormous number of potentially toxic chemicals. Among them are the following:

- *Alpha-terpineol* is highly irritating to mucous membranes. It can also cause excitement, ataxia (loss of muscular co-ordination), hypothermia, CNS* and respiratory depression, and headache. Repeated or prolonged skin contact is not advised.

- *Benzyl acetate* vapours are irritating to eyes and respiratory passages, and may cause coughing. The chemical can be absorbed through the skin, causing systemic effects. It is a carcinogen that has been linked to pancreatic cancer.

- *Benzyl alcohol* is irritating to the upper respiratory tract. It can cause headache, nausea, vomiting, dizziness, drop in blood pressure, CNS* depression and, in rare cases death due to respiratory failure.

- *Camphor* is on the EPA's hazardous waste list. It is readily absorbed through body tissues; vapours and inhalation can irritate the eyes, nose, and throat. It is a CNS* stimulant that is associated with dizziness, confusion, nausea, twitching muscles and convulsions.

- *Chloroform* is on the EPA's hazardous waste list. It is a neurotoxin, anaesthetic and carcinogen. Inhalation of vapours may cause headache, nausea, vomiting, dizziness, drowsiness, irritation of respiratory tract and loss of consciousness. Chronic effects of over-exposure may include kidney and/or liver damage. Its use can aggravate medical conditions such as kidney disorders, liver disorders, heart disorders and skin disorders.

- *Ethyl acetate* is on the EPA's hazardous waste list. It can be irritating to the eyes and respiratory tract; it is a narcotic and exposure may cause headache and narcosis (stupor). It has also been linked to anaemia with leukocytosis (an increase in the number of white blood cells; this is an immune system reaction) and damage to liver and kidneys.

- *Limonene* is an irritant to the skin and eyes and may cause allergic reactions. It is also a carcinogen.

- *Linalool* is a narcotic. It has been shown to cause CNS* disorders. Animal tests have proven it to adversely affect mood and muscular co-ordination and reduce spontaneous motor activity. Exposure can cause fatal respiratory disturbances. It even attracts bees (thus posing a threat to people who are allergic to bee stings).

- *Pentane* is irritating to the eyes and can cause skin rashes. Its vapours may cause headache, nausea, vomiting, dizziness, drowsiness, irritation of respiratory tract and loss of consciousness. Prolonged and repeated inhalation of vapours may cause CNS* depression.

   *CNS is your central nervous system (i.e. you brain and your spine). CNS disorders include Alzheimer's disease, attention deficit disorder (ADD or ADHD), dementia, multiple chemical sensitivity (MCS), multiple sclerosis (MS), Parkinson's disease, seizures, strokes, sudden infant death syndrome (SIDS). CNS exposure symptoms include: aphasia, blurred vision, disorientation, dizziness, headaches, hunger, memory loss, numbness in the face, and pain in the neck and spine.

Alternatives

Nearly all commercial brands of laundry detergent and fabric softener leave chemical residues on your clothes that can irritate the skin. And while most are made to biodegrade within a matter of days, their manufacturing process is far from ecological. For a healthier environment buy washing powders and liquids that use only vegetable-based detergents. Also for everyday stains, non-biological detergents are perfectly adequate. Try to avoid fabric softeners altogether. Or try other laundry alternatives:

- *Choose powders* over liquids and since we all tend to use more detergent than necessary to wash clothes, use less of whichever type you buy.

- *Try pure soap flakes instead of detergent.* Add ½ cup of borax or vinegar to the rinse to remove soapy residue. Vinegar has the advantage of acting like a fabric softener and mould inhibitor as well.

- *To remove heavy soils* add 1 teaspoon of trisodiumphosphate (TSP) to the wash. Borax and washing soda (sodium carbonate) can also remove heavy stains.

- *To remove perspiration* and odours pre-treat with baking soda, white vinegar or borax dissolved in water. These ingredients can also be added to the wash to freshen the load.

- *Make use of a washing line* if you have one. The sun's rays are the best bleaching agent in the world. Clothes that have been gently dried in the sun also tend to be naturally softer than those tumbled in the drier.

Laundry Boosters

Most laundry detergents have a small amount of added bleach, but not enough to handle really tough stains. This is why there are now separate bleaches that you can add to the wash to boost cleaning power. Laundry boosters used to be made with common bleach (sodium hypochlorite) – the same ingredient used in swimming pools and drain cleaners.

However, as fabrics have become more colourful and sensitive to the effects of common bleach, manufacturers have come up with new colour-sparing bleaches based on sodium perborate or sodium chloride. Sodium perborate breaks down into hydrogen peroxide in the wash and liberates oxygen which oxidises, or bleaches. Sodium chloride liberates chlorine gas to oxidise or bleach stains. Chlorine is the more powerful of the two types of bleach; it is also the more irritating to skin, eyes and lungs. All bleaches are corrosive and harmful if swallowed.

Pre-treatments and Stain Removers

These are available in a variety of forms including pump sprays, liquids, gels, solids, or aerosol. They contain surfactants and solvents. Aerosol versions contain a variety of petroleum distillates and propellants. Most pre-treatments are just liquid detergents under another name. Logic dictates that if you put the detergent in high concentration on a stain before you start to wash you have a better chance of removing that stain.

Solid or stick formulas are probably best for really ground-in stains because the action of rubbing them on to the fabric ensures that they are pushed deeply into the stain. They also contain fewer chemicals. Frankly, you can substitute a liquid soap or detergent for most pre-treatments and it will do exactly the same job.

Getting Starched

Not many of us starch in our laundry these days. Good thing too. Laundry starches, while based on corn starch formulas, contain a number of nasties including *formaldehyde* – a carcinogen and a strong irritant to the eyes, throat, skin and lungs; *phenols* – carcinogens which also cause central nervous system depression, can severely affect the circulatory system, and are corrosive to skin; and *pentachlorophenol* – a reproductive toxin associated with foetal abnormalities and birth defects.

Dry Cleaning

When you use a home dry-cleaning fluid or send your clothes to be dry-cleaned you are soaking your clothes in a mixture of some of the most hazardous chemicals around. Not surprisingly, there is a large body of medical evidence showing that people who work in the dry-cleaning industry suffer from a wide range of health problems including kidney and liver disease, cancer of the larynx and oesophagus, optic neuritis (an inflammation of the optic nerve resulting in diminished eyesight) and memory impairment. The ingredients in dry-cleaning fluids, whether used at home or professionally, are basically the same. Typically they include the following:

- *Carbon tetrachloride* is a toxic solvent that produces cellular destruction throughout the body, especially in the liver, kidney, and central nervous system. It is toxic by all routes of exposure: inhalation, absorption, skin contact, and oral ingestion. In 1970 the FDA classified carbon tetrachloride as a substance so hazardous that no warning label could be devised that would adequately protect the householder. It was subsequently banned from use in household products though it remains in use in industry and as a fumigant.

- *Perchloroethylene*, also known as tetrachloroethylene, ethylene tetrachloride, or PERC, is a carcinogen. It is fat-soluble and so collects in the tissues of living organisms and accumulates in the environment. Its vapours are irritating to skin, eyes, and upper respiratory tract and can produce giddiness, headache, inebriation, nausea, vomiting, and sinus inflammation. Chronic inhalation can cause serious depression of the central nervous system as well as kidney and liver damage. The coin-operated dry-cleaning machines found in laundrettes have been associated with acute tetrachloroethylene poisoning.

- *Trichloroethane* is commonly used as a solvent and cleaning agent in spot removers and fabric cleaners, film cleaner, insecticides, paint and varnish remover, degreaser, typewriter correction fluid, and as an aerosol propellant. It is absorbed by inhalation and ingestion, is an irritant to the eyes and nose and can result in central nervous depression and liver and kidney damage if ingested.

- *Naphthas* are derived from both petroleum distillation and coal tar. Petroleum naphtha has a lower order of toxicity than coal tar naphtha. However, over-exposure to either type may cause central nervous system depression, with symptoms of inebriation followed by headache and nausea. Naphthas can enter into your system through inhalation and vapours, ingestion, and eye and skin contact. They are irritating to the skin, eyes, and upper respiratory tract. Skin chapping and sensitivity to light may develop after repeated contact.

- *Benzene* is a carcinogenic petroleum distillate. It is poisonous and irritating to mucous membranes. Inhalation of fumes can be acutely or chronically toxic. Harmful amounts may be absorbed through the skin and may cause sensitivity to light, and produce skin rashes and swelling. For over a century, scientists have known that benzene is a powerful bone marrow poison, destroying the bone marrow's ability to produce blood cells.

- *Toluene* is also known as toluol or methylbenzene. It and its chemical cousin xylene (also known as xylol or dimethylbenzene) are aromatic hydrocarbons used primarily as solvents. They easily enter your system through inhalation and ingestion, but are poorly absorbed by the skin. Once inhaled they are narcotic and are irritating to the skin and respiratory tract. They can cause damage to several organs including eyes, liver, kidneys, and skin as well as to the central nervous system. Symptoms of chronic exposure include fatigue, weakness, confusion, headache, watery eyes, muscular fatigue, insomnia, dermatitis, and photosensitivity.

Alternatives

Take a sniff of your clothes when they come back from the dry cleaners. Most dry cleaning machines are built to recapture the chemicals used in the cleaning process so that they can be used over and over again. But if you can smell any chemicals at all on your clothes it may be time to change dry cleaners. To minimise the risk:

- *Buy clothes that do not need dry cleaning.* Dry cleaning is a luxury that should be reserved for the most precious garments. Often it is simply unnecessary and manufacturers recommend dry cleaning in order to cover themselves against any potential damage done through careless washing.

- When you bring clothes home from the dry cleaners remove the plastic bag and *allow the clothes to air* near an open window before wearing.

- If you are using a commercial household spot remover always *wear rubber gloves* and work so that the fumes are blowing away from you. Do not allow children or pets in the room when you are working. Keep the lid on the product while in use to avoid solvent being vaporised in the room. Always wash splashes on the skin immediately with soap and plenty of water.

- *Never use dry-cleaning fluid in a washing machine* or put articles that are damp with dry-cleaning fluid in the dryer because most dry-cleaning fluids are flammable; those that are not will produce toxic gasses in the dryer.

These days you can buy special sheets that claim to freshen dry-cleanable clothes in the dryer. Approach these with caution. They contain powerful spot removers similar to home dry-cleaning fluids and strong surfactants and perfumes both of which can cause allergic reactions.

Chapter 19

Oven Cleaners and Drain Cleaners

Down the Drain

'Why do oven cleaners and drain cleaners appear together?', you might ask. Because they are both made from the same strongly caustic ingredients.

There are two basic types of drain cleaners. Those used for maintenance are called build-up removers. They contain enzymes, or cultured bacteria that produce enzymes, and are formulated to dissolve grease and soap scum.

Conventional drain cleaners, or drain openers, are highly corrosive and dangerous to use. This is because their main ingredients are *sodium hydroxide* (lye), a caustic that can cause burns to the skin and, in severe cases, blindness, and *sulphuric acid*, a corrosive chemical that can also cause severe skin burns and blindness. Both work by eating away at whatever is in their path. This includes your skin if you are unlucky enough to splash some on yourself. The vapours from commercial drain cleaners are also harmful.

The use of chemical drain cleaners as a preventative is not a good idea. If you have a septic tank be aware that chemical drain openers can kill off all the beneficial bacteria necessary for the efficient working of the tank.

Even build-up removers that claim to be 'non-corrosive' or 'non-caustic' can contain chemicals that can be poisonous if inhaled or swallowed. Such products should clearly state their ingredients.

Alternatives

Regularly running boiling water down the drain is probably the best way to keep drains in order. Since greasy deposits are a major cause of drain blockages, try not to pour grease down the drain if you can help it. In addition:

- *Make a foaming drain cleaner* out of 200g baking soda, 100g table salt and 200g vinegar. When the alkali (baking soda) meets the acid (vinegar) they will bubble furiously, pushing the abrasive salt through the drain. All three ingredients will work to remove deposits from the drain and clear it. After 20 minutes or so pour boiling water down the drain to clean it out.

- *Use a drain strainer.* These little filters can be used in bathrooms and kitchens to keep large particles from going down the drain and forming blockages.

- *Use a plunger.* This can be a very effective way of dislodging blockages. Once you have opened the drain, use boiling water or the foaming cleaner above to clear out the drain. Be warned though that you should not use a plunger after using chemical drain openers. This can invite splashback on to the skin and into the eyes.

Never mix commercial drain openers with anything else. Not even natural alternatives. The combination could become reactive and blow out of the sink and on to you.

Oven Cleaners

The majority of oven cleaners come in aerosol containers. This means that in addition to lye and sulphuric acid you are also being exposed to propellants. The reason why oven cleaners are made with lye is simple. When the lye mixes with grease in your oven it makes crude soap that can then be used to facilitate cleaning.

Strong alkalis can be used on cold ovens. However, there are newer types of oven cleaners that use less alkali chemicals. These need to be used on warm ovens to aid removal of grease and grime. The vapours from both types are nevertheless highly toxic.

Alternatives

Prevent messy grime from building up by placing a sheet of aluminium foil on the oven floor. This will catch most spills and can be changed as often as necessary.

- *Use baking soda.* This can be sprinkled into the still warm oven to loosen grease and then wiped away using warm water. Or you can make a spray from 3 tablespoons of baking soda dissolved in a pint of warm water. Spray on and wait 20 minutes. Scrub off with a fine wool pad if necessary. Baking soda is effective for light- to medium-soiled ovens. Heavily soiled ovens may require a different approach.

- *Use vinegar.* Using the same principles for unblocking a drain you can spray a solution of white vinegar into your warm oven. Sprinkle with baking soda. Leave to bubble and then scrub.

Chapter 20

Polishes

Shine at a Price

Nothing says clean like a nice sheen on your furniture, metals, shoes and car. But to achieve that nice sheen we use a mixture of dangerous solvents and waxes derived from petroleum.

Most polishes are flammable and all contain hazardous ingredients such as solvents, petroleum distillates, volatile organic compounds, formaldehyde and benzene. The health dangers most often associated with polishes of all kinds are inhalation of fumes or vapours (especially from aerosols) and poisoning from ingestion. Polishes that look drinkable, like strawberry soda or milk, are especially tempting to children.

Furniture Polish

There are three basic types of commercial furniture polish, each of which uses a different type of chemical to aid the application of the wax or oil to the furniture surface. *Solvent* polishes use a chemical solvent to dissolve the oil or wax into a liquid form. *Emulsion* polishes suspend the wax in a liquid, usually water. *Aerosol sprays* are solvents or emulsion types packed under pressure.

Furniture polish (and other types of polish as well) may contain one or more of the following substances:

- *Naphthas* are toxic solvents derived from both petroleum distillation and coal tar. They can enter your system through inhalation or be absorbed through the skin. Skin chapping and sensitivity to light may develop after repeated contact. Over-exposure may cause central nervous system depression with symptoms of drunkenness. When using a product that contains naphthas, be certain to wear gloves and make sure there is adequate ventilation.

- *Nitrobenzene* is a highly toxic solvent that can cause spleen and liver damage. It can enter your body through inhalation, skin and eye contact, and ingestion. It affects the central nervous system, producing headache, fatigue, general weakness, vertigo and in some cases severe depression, unconsciousness, and coma. Drinking alcohol increases the toxic effects of nitrobenzene. Nitrobenzene quickly crosses the placenta and can cause birth defects. Products containing nitrobenzene should be avoided.

- *Phenol*, also known as carbolic acid, is flammable, corrosive, and very poisonous. Light sensitivity and sinus congestion are common with exposure to phenol fluids or vapours. Phenol and related compounds rapidly strip protective oils from the skin and can cause severe burns, skin ulcerations, skin rashes, swelling, pimples, and hives. Fatal poisoning can also occur through skin absorption. Because it can act like a local anaesthetic, users may not notice the damage to the skin until it becomes apparent to the eye. Ingestion of even small amounts may cause vomiting, convulsions, circulatory collapse, paralysis, and coma.

- *Ammonia* liquid can be corrosive, causing severe burns and irritation to the skin, eyes, and lungs. Vapours, even in low concentrations, can cause severe eye, lung, and skin irritation. Chronic irritation may occur if ammonia is used over long periods of time.

- *Petroleum distillates.* These refer to a wide range of compounds that are a by-product of distillation during the refining of crude oil. Always wear gloves to avoid skin contact and avoid breathing vapours of volatile compounds. In addition, keep petroleum distillate products out of the reach of children. Do not mix different petroleum distillate products.

Alternatives

Caring for wood is really very simple. Often it comes down to how you treat the wood in between cleans, rather than the polish that you buy. Do not let hot or wet things stand on the surface of the wood. Protect your furniture from hard knocks and sharp points like pens and it should be able to stand up to everything else that everyday life throws at it.

Although wood originally came from a living thing, wooden furniture is not alive. It does not need 'nourishing' or 'feeding'. The purpose of polish is mainly to make the wood look nice, and to keep natural moisture from getting out. If your furniture was varnished or polished in the factory you only need to give it a good going-over once or twice a year. The rest of the time a simple dusting is sufficient. Over-use of aerosol sprays in particular can leave a slightly milky film on polished wood which will be impossible to get out except by stripping and refinishing the wood.

Unvarnished wood can be cleaned and polished easily and beautifully with very simple compounds. For instance, a dab of vinegar on a slightly damp cloth can be a very effective cleaner. A little light olive oil on a cloth will also polish and protect your wood. However, if you are feeling more ambitious try making the following alternatives:

- *Simple furniture polish.* Fill a small bottle with two-thirds oil of your choice (olive and walnut are good choices or a mixture of one of these with linseed oil in a ratio of 2:1). Top up the bottle with white vinegar and then add 20–30 drops of essential oil (lemon and bergamot are good choices). Shake well before each use and apply sparingly with a soft cloth. Use a second clean cloth for buffing (this is an essential step to avoid build-up). The vinegar will help clean and the oils will protect the wood.

- *Furniture wax.* Use this mixture occasionally to polish and protect wood.

  2 tblsp beeswax
  3 tblsp oil
  1 tblsp white vinegar.

Grate the beeswax into a small ceramic bowl or glass-measuring cup (this helps speed the melting process). Add the oil and vinegar and put the bowl into a shallow pan of boiling water. Stir it as it mixes and when it has fully melted leave to cool a bit. While the mixture is still liquid pour it into a small jar. Apply sparingly and leave to dry before buffing with a clean cloth.

Metal Polish

Most metal polishes contain strong acids such as phosphoric acid or oxalic acid. They can also contain caustics such as ammonia hydroxide, and reproductive toxins and neurotoxins such as petroleum distillates, nitrobenzene, naphthalene and ethanol.

The most dangerous polishes contain sulphuric acid and hydrofluoric acid. The hydrofluoric acid found in some rust removers and aluminium polishes can eat right through the skin and down to the bone. Less toxic polishes contain detergents, trisodium phosphates and pine oils. Unfortunately these also tend to be the least effective when it comes to removing tarnish.

Metal polishes are acidic because acid dissolves tarnish. Unfortunately they can sometimes be so harsh that they damage the item you are trying to clean. They can eat through the metal or metal plating. They can destroy the lacquer coatings commonly used on items such as brassware and leave permanent stains on the surface of every type of metal. This is especially true of the quick 'dip' type polishes.

Alternatives

Combinations of natural acids such as lemon juice or vinegar and salt make good metal cleaners. However, any chloride such as salt or washing soda can damage fine metals so do not use these home-made polishes on family heirlooms. If a piece is of particular value, you might consider having it professionally cleaned.

Brass and copper

Rub the pans with a mixture of salt and vinegar or lemon juice (or a lemon rind soaked in salt) and then wash. Heating the vinegar will aid the process since all acids are more active when hot. Give this time to work. The best way is to work the mixture all over the brass or copper item and then let it sit for a while.

Alternatively put 15g (½ oz) citric acid in ½ litre (1 pint) of water. You may need to adjust the amount of acid by trial and error. Swab this on your metal item and leave for a little while. Make sure you place the item on plenty of newspaper or other protective covering and leave a window open. This will remove tarnish and even rust, making polishing much easier.

Bronze

Bronze should not be polished. It can easily be damaged by water and polishing will remove the patination that is part of its character. Brush very occasionally with a clean soft brush and use a cotton bud to remove dirt from crevices.

Pewter

Likewise pewter is easily damaged by metal polishes. The best way to protect it is by keeping it dust free with a clean dry cloth and occasionally rubbing the outside with a silver cleaning cloth. Never rub polish on the inside of a pewter goblet or any other vessel intended to hold liquid. Pewter is absorbant and this can make it unpleasant and even dangerous to drink from.

Silver

Some experts believe that a bit of tarnish in the crevices brings out the character and design of the piece. So do not over-polish your silver. Intricate designs are best brushed with a soft brush to get the polish into the detail.

You can make your own simple silver dip by putting a plate-sized sheet of aluminium foil in a saucepan with water and a handful of washing soda. Heat this and place the silver in the pan for a minute or two, until the tarnish and stains are just removed. *Do not* leave for a long time and *do not* use this method for cleaning silver plate. Buff with a soft, clean cloth or long-term silver cloth.

To clean the inside of a silver teapot fill it with boiling water and add 1 teaspoon of denture cleaner and leave overnight.

Silver plate should only be polished when absolutely necessary. Try using a special silver polishing cloth for this job and remember to wear gloves.

Whiting, a fine powder that can be purchased at most hardware stores, is useful for polishing silver. To make your own silver polishing cloth mix 4 teaspoons of whiting with 2 tablespoons of household ammonia and 3 cups of water. Soak a soft absorbent cloth in this, wring it out and let it half dry (preferably near an open window), then store in a resealable plastic bag for use as needed.

Chrome
Rather than a commercial mixture of ammonia and other chemicals, why not try straight ammonia and water? Put a dash of it in water and dip a soft cloth in the mix. Rub this on the chrome, rinse and then polish with a clean, soft cloth. Do wear protective gloves and try not to breathe the fumes, and if you are washing the car be careful not to get the ammonia on the paintwork. Chrome fixtures in the bathroom require only soap and water. Stubborn stains can be lifted by rubbing gently with a bit of bicarbonate of soda on a slightly damp soft cloth.

Stainless steel
Even everyday stainless steel pans, dishes and cutlery can be damaged by too much polish and scratchy sponge cleaners. To clean stainless steel you only need hot water and a little detergent. If you really feel the need of a stronger mixture use a dash of ammonia in hot water. Drying the items immediately will prevent annoying spots from forming.

Marble Polish

Marble gives the impression of strength and grandeur but it is really rather soft and easy to chip, break and stain.

Marble should never be cleaned with spirits or acids. Even water will eventually stain and weaken marble so it is best dealt with by using a dry, soft cloth. If you absolutely have to clean marble try doing it with a little white spirit and a clean cloth. Family heirlooms should be cleaned by experts. There is no way to avoid cleaning marble floors so try using water and weak detergent solution. If possible wash and then thoroughly dry the floor to prevent excess moisture seeping into the stone.

Shoe Polish

It is unfortunate that most shoe polishes do not list their ingredients. Many commercial shoe polishes contain a number of suspected human carcinogens that can be easily absorbed through your skin. Among them:

- *Trichloroethylene*, a toxic solvent and degreaser. It is commonly used in septic tank cleaners. Trichloroethylene fumes and residues can linger on surfaces for a long time where they can produce central nervous system symptoms such as headache, fatigue and confusion.

- *Nitrobenzene*, a highly toxic substance (see details under furniture polish).

- *Methylene chloride*, also known as methylene dichloride and dichloromethane. Fumes can produce symptoms similar to carbon monoxide toxicity. Memory loss, skin irritation, liver and kidney damage are reported with chronic exposure. Methylene chloride is a known animal carcinogen and a suspected human carcinogen. When heated, it emits a highly toxic phosgene gas (nerve gas). The use of products containing methylene chloride by people with heart conditions has resulted in fatal heart attacks.

There is no doubt that shoes last longer and look better when they are properly taken care of. What is more, very few good alternatives to conventional shoe polish exist. Nevertheless, while most of us would never think that there are safety issues connected with shoe polishing, there are. To protect yourself from the nasty chemicals in shoe polish:

- *Wear gloves* when polishing or cleaning shoes – the chemicals in the polish are easily absorbed into your skin.

- *Make sure shoes are dry before wearing.* Many of the vapours emitted from shoe polish interact with alcohol. The presence of alcohol in the system, for instance, heightens the toxic effects of nitrobenzene. For this reason it is wise not to wear shoes that are still wet with polish if you are going out (or staying in) drinking.

- *Keep shoe polish out of reach of children.*

Polishing the car?

Car waxes typically contains 75–85 per cent petroleum naphtha and 15–25 per cent wax. Naphtha is flammable and an irritant that can enter your system through inhalation, ingestion, and skin and eye contact. Skin chapping and sensitivity to light may develop with repeated and prolonged contact. When you are polishing the car, wear protective gloves. When the polish is not in use, make sure that the top of the tin or bottle is securely closed and that the bottle is kept out of the reach of children.

Postscript

The purpose of this book is not to make you panic. It is not to encourage you to clear the bathroom and kitchen cupboards of every product you have ever bought. Nor is it to suggest that every product you use is going to make you sick.

The purpose of this book is to provide information and to help you, as a consumer, become more aware of the types of chemicals you come into contact with each day and become more selective about the products you buy.

With this knowledge you can begin the task of taking better care of yourself and your family. The chemicals listed in this book are only a small part of the thousands of chemicals we come into contact with every day in our homes. These in turn are only a fraction of the tens of thousands of chemicals we come into contact with in daily life. But the good news is that, unlike outdoor pollution, you can do something about the chemicals present in your home.

Doing this is good for your body, because it gives you a mandate to begin reducing the toxic burden on your system. The benefits you (and especially your children) derive from this are likely to be improvements in your short- and long-term health. Personally I am a firm believer in the old axiom that when you have your health, you have everything.

But taking control of the chemicals in your home is also good for your mind and your soul. Being familiar with the chemicals in everyday products puts you back in control of what you spend your money on. It also gives you a certain amount of control over your own destiny.

Unless you are very motivated, there is no need to make swinging changes in your life. Even small changes in the way you choose and use personal care or household products will be a step in the right direction. But as each product runs out why not consider replacing it with something less toxic? Enjoy experimenting with alternatives and finding what works for you.

For some, choosing to use a number of the home-made alternatives in this book may prove to be a strangely liberating experience. Suddenly you are self-sufficient and your well-being does not depend on the latest list of products from some anonymous corporation.

What this book has done is give you some basic building blocks. What you now choose to build into your life with them is entirely up to you.

Glossary of Terms

Acid A *corrosive* substance with a low pH (1–7). Acids turn blue litmus paper red. Strong acids can dissolve substances like grease but can also damage skin and tissue. Acids neutralise *alkalines*.

Acute effect An effect which occurs shortly after (usually within 24 hours) continuous short-term exposure to a chemical.

Alkaline Also called base, is a substance that is *caustic* and has a high pH (7–14) and is used to neutralise acids. Turns red litmus paper blue.

Blood brain barrier The protective membrane that surrounds the brain. The purpose of the blood brain barrier is to prevent the accumulation of toxins and other harmful substances in the brain. Many common chemicals, particularly fragrances, can breach the blood brain barrier and this membrane is not fully formed in young children, who are particularly at risk from such substances.

Carcinogen A substance that can cause cancer.

Caustic A highly *alkaline* or highly *acid* substance capable of burning, irritating and destroying skin and tissue.

Chronic effect Symptoms that can be mild or severe (or anything in between); appear gradually over time with continual low-grade exposure to a chemical.

Corrosive A substance capable of dissolving what it comes into contact with, be it man-made materials or living tissue. Both *acids* and *alkalis* can be corrosive and also poisonous if ingested.

Flammable A substance that will ignite or burn easily when exposed to heat. Most flammable chemicals are also toxic.

Hard water Water containing calcium and magnesium salts that have dissolved from the rocks over which the water has flowed. Water without these salts is called soft water.

Hazardous A warning that prolonged exposure to the product is toxic and capable of causing illness, damage or even death.

Immunotoxin A substance that reduces the effectiveness of the immune system, leaving the body defenceless against disease.

Irritant Anything capable of causing a local reaction in the body. Examples of irritant reactions include skin rash or inflammation, itching, respiratory difficulties or watery eyes.

Local effect Something that affects only a single organ or tissue. Often used to describe skin reactions to chemicals but can also refer to chemicals that target specific internal organs.

Mutagen A substance capable of causing mutations in cellular DNA. If this change occurs in a developing sex cell the mutation can be passed on to offspring. Mutagens can also cause cancer.

Neurotoxin A chemical that is toxic to the nervous system. The nervous system includes the brain, spinal cord, nerves and sensory receptors. Nearly all bodily functions are controlled in some way by the nervous system. Symptoms of neurotoxicity include lowered IQ, poor short-term memory, numbness in the extremities, headache, loss of concentration, blurred vision and slowed reaction time. Long-term exposure to neurotoxins can also cause cancer.

pH Term used to describe the acidity or alkalinity (base) of a substance. Substances with a pH of 1–7 are considered acids while those with a pH of 7–14 are considered alkaline. Acids and bases neutralise each other. The scale is 'logarithmic', which means, in this case, that pH 3 is considered ten times more acidic than pH 4.

Poisonous A substance that is highly toxic even when ingested in small amounts and that may even be capable of causing death.

Sensitiser A substance that produces symptoms in certain individuals even at low doses. People who are affected by one sensitiser can become increasingly sensitive to other chemicals with which they come into contact.

Surfactant A family of chemicals, which includes detergents, often found in cleaning products. They work mainly by changing the way water interacts with other chemicals. Surfactants break down the surface tension of water, giving it a greater wetting ability. They also aid the emulsifying, foaming, and dispersing properties of water.

Surfactants are either amphoteric, non-ionic, cationic or anionic, according to their magnetic charge. *Amphoteric* surfactants are both positively and negatively charged; *non-ionic* surfactants are neither positively nor negatively charged. *Anionic* surfactants are negatively charged and *cationic* surfactants are positively charged. The magnetic charge of a surfactant determines its use. Amphoterics are widely used in personal care products of all kinds. Non-ionics are used in detergents meant to cut through grease (i.e. on the hair or face or in the kitchen). Anionics are often used in high foaming detergents such as shampoos, foam baths and dish detergent. Cationics are used mostly in disinfectants and sanitisers, but also in hair conditioners. Cationics have the greatest ability to penetrate the skin.

Synergistic effect An effect that is much greater than might be predicted from the combination of individual chemicals. When chemicals combine synergistically they enhance each other's effectiveness, not always to the good. An example is the comination of alcohol and barbiturates in the stomach.

Systemic effect Something that affects the body as a whole organism rather than targeting a single specific organ or tissue.

Teratogen A substance that causes birth defects.

Toxic A substance capable of causing injury or death when it enters the body either through ingestion, inhalation or skin absorption.

Volatile Liquid chemicals that readily convert into gases under ordinary (for instance room temperature) conditions and thus can rapidly be dispersed in the atmosphere and easily enter the body via the lungs.

Useful Contacts

Below is a short list of firms that specialise in producing less toxic products. With a rare few exceptions it is impossible to unreservedly recommend every product that each of these companies produces. Instead it is advised that consumers ask questions, read labels and choose for themselves those products which they feel safe with.

Toiletries

Aveda produces upmarket toiletries and cosmetics that also contain some herbs and organic ingredients and are available in department stores throughout the UK (phone 0800 328–0849 for details) or on the net from www.aveda.com

Desert Essence products, which contain some organic essences and ingredients, are available from healthfood stores or by mail order from Revital Stores at www.revital.com

Dr Bronners castile soap, which can be used on body and in household cleaning, can be ordered from: 21st Century Health Ltd, 3 Water Gardens, Stanmore, Middlesex HA7 3BR. Tel: 020 8420–6467; web: www.21stcentury.co.uk

Green People make toiletries for the whole family, which include some organic essences and ingredients. They are at: Brighton Road Handcross, Haywards Heath, West Sussex RH17 6BZ, Tel: 01444 401444; web: www.greenpeople.co.uk

Neal's Yard Remedies produce a range which includes real soap, herbs and essential oils. They have shops throughout the UK. For details of their mail order catalogue call 0161–831–7875. Or contact the shops closest to you:

Brighton: 01273 601464
Bristol: 0117 946 6034
Bromley: 020 8313 9898
Cardiff : 01222 235721
Cheltenham: 01242 522136
Edinburgh: 0131 226 3223
London Camden: 020 7284 2039
London Chelsea Farmers Market: 020 7351 6380
London Covent Garden: 020 7379 7222
London W11: 020 7727 3998
Manchester: 0161 835 1713
Newcastle upon Tyne: 0191 232 2525
Norwich: 01603–766681
Oxford: 01865–245436

Weleda produce a wide range of products many of which come highly recommended. These are available in healthfood shops or by mail order at their website on www.weleda.co.uk

Household
Bio D a major UK manufacturer of environmentally-friendly cleaning products such as washing powder, dish detergent, multi-surface cleaner and toilet cleaner. Can be found in many independent healthfood stores or you can order by phone from Bio D direct on: 01482 229950 (minimum purchase £25).

Ecover make a wide range of environmentally-friendly, low-toxic household products. They are available in some supermarkets, healthfood shops and by mail order from Ecover Direct on 01635–528240

Faith in Nature produce a range of cleaning products that are low in toxic chemicals. They are located at: Unit 5, Kay Street, Bury, Lancs, Tel: 0161 764 2555

Laundry balls are an alternative to powders and claim to reduce surface tension and 'ionise' the water (laundry powders do much the same thing). They are not totally non-toxic but they can be economical for washing lightly- to medium-soiled clothes. Instead of foaming detergents they usually contain highly compressed beads of alkyl-based surfactants and washing soda in a large plastic ball. The beads look like sweets so keep out of reach of children. Two popular brands: **Ecoballs** can be purchased by mail order on: 020 8777–3121 or by visiting their website at www.eco-ball.co.uk also at health food shops. **Aqua Balls** are available through 21st Century Health Ltd, 3 Water Gardens, Stanmore, Middlesex HA7 3BR. Tel 020 8420–6467; web: www.21stcentury.co.uk

Soapnut is an organically cultivated multipurpose cleaning powder, cultivated and used for centuries in India. You can use it on clothes and in the bath. For more information on soapnut products phone: 01284 700 170 or visit their website on www.soapnut.com

Organisations

British Society for Allergy, Environmental and Nutritional Medicine (BSAENM)
c/o The Burwood Clinic
34 Brighton Road
Banstead
Surrey SM17 1BS

Can put you in touch with an experienced GP local to you who specialises in allergies.

Cosmetic Toiletry & Perfumery Association (CTPA) Limited
Josaron House
5/7 John Princes Street
London W1G 0JN
Tel: 020 7491 8891
Fax: 020 7493 8061
email: info@ctpa.org.uk

The CTPA is an umbrella organisation that looks after the interests of cosmetic, toiletry and perfume manufacturers.

Friends of the Earth
26–28 Underwood Street
London N1 7JQ
Tel: 020 7253–4248

FoE produces many reports and other data on the use of chemicals, which consumers may find interesting.

Greenpeace
Canonbury Villas
London N1 2PN
Tel: 020 7354–5100

A campaigning organisation that produces consumer-friendly information on, among other things, toxic exposures.

Women's Environmental Network
87 Worship Street
London EC2A 2BE
Tel: 020 7247 3327

A campaigning group concerned with a variety of environmental issues that affect mothers and babies.

In Emergencies

The UK National Poisons Information Service (NPIS) is made up of six centres throughout the UK. It maintains links with the poisons unit in Beaumont Hospital in Dublin, and the wider toxicology community, through the European Association of Poisons Centres and Clinical Toxicologists and the American Academy of Clinical Toxicology.

If your child has swallowed any household toxin you can also phone the following numbers for help and advice:

National number (for all poisons enquiries): 0870 600 6266

Local centres

Belfast: 01232 240503
Birmingham: 0121 507 5588/9
Cardiff: 01222 709901
Dublin: 00353 1 8379964
Edinburgh: 0131 536 2300
London: 020 6635 9191
Newcastle: 0191 232 5131

Select Bibliography

An enormous amount of research has formed the background to this book. These pages represent a selection of the papers and books that readers may find interesting.

Papers and Reports

'A case-control study of borderline ovarian tumors: the influence of perineal exposure to talc', Harlow, B.L. and Weiss, B.S., *American Journal of Epidemiology*, 1989; 130: 390–4

'A case-control study of hair dye use and breast cancer', Shore, R.E. et al, *Journal of the National Cancer Institute*, 1979; 62: 277–83

'A prospective study of permanent hair dye use and hematopoietic cancer', Grodstein, F. et al, *Journal of the National Cancer Institute*, 1994; 86: 1466–70

'An environmental assessment of alkylphenol ethoxylates and alkylphenols', Warhurst, A.M., Friends of the Earth, London, 1995

'Chemical hazard data availability study', US Environmental Protection Agency, 1998

'Crisis in chemicals: The threat posed by the "biomedical revolution" to the profits, liabilities and regulation of industries making and using chemicals', Friends of the Earth, May 2000

'Environmental and heritable factors in the causation of cancer – analyses of cohorts of twins from Sweden, Denmark, and Finland', Lichtenstein, P. et al, N Eng *Journal of Medicine*, 2000; 343: 78–85

'Environmental medicine, part 1: the human burden of environmental toxins and their common health effects', Crinnion, W.J., *Alternative Medical Review*, 2000; 5: 52–63

'Everyday exposure to toxic pollutants', Ott, W.R. and Roberts, J.W., *Scientific American*, February 1998

'Genital talc exposure and risk of ovarian cancer', Cramer, D.W. et al, *International Journal on Cancer*, 1999; 81: 351–6

'Hair dye use and multiple myeloma in white men', Brown, L.M. et al, *American Journal on Public Health*, 1992; 82: 1673–4

'Hair product use and the risk of breast cancer in young women', Cook, L.S. et al, *Cancer Causes Control*, 1999; 10: 551–9

'Have you met this syndrome?', Dr Richard Lawson, *Medical Monitor*, September 4, 1996: 66

'Health hazard information', US Environmental Protection Agency, 1991

'Identification of polar organic compounds found in consumer products and their toxicological properties', Cooper, S.D. et al, *Analysis of Environmental Epidemiology*, 1995; 5: 57–75

'Identification of polar volatile organic compounds in consumer products and common microenvironments', US Environmental Protection Agency, 1991

'Increasing incidence of non-Hodgkin's lymphoma: occupational and environmental factors', Pearce, N. and Bethwait, P., *Cancer Research*, 1992; 52(19 Suppl): 5496s-5500s

'Multiple chemical sensitivity recognition and management: A document on the health effects of everyday chemical exposures and their implications', Third scientific report of the British Society for Allergy, Environmental and Nutritional Medicine, Eaton, K.K., Anthony H.M., *BSAENM*, March 2000

'Mutagenicity of cosmetic products containing Kathon', Connor, T.H. et al, *Environ Mol Mutagen*, 1996; 28: 127–32

'Seventh annual report on carcinogens', National Toxicology Program, 1994 US Department of Health and Human Services

'Neurogenic inflammation and sensitivity to environmental chemicals', Meggs, W.J., *Environmental Health Perspectives*, 1993; 101: 234–8

'Neurogenic switching: A hypothesis for a mechanism for shifting the site of inflammation in allergy and chemicals sensitivity', Meggs, W.J., *Environmental Health Perspectives*, 1995; 103: 54–6

'Neurotoxins: at home and the workplace', report by the Committee on Science and Technology, US House of Representatives, Sept 15, 1986, No 99–827

'N-Nitrosoalkanolamines in cosmetics in relevance to human cancer of N-nitroso compounds, tobacco smoke and mycotoxins', Eisenbrand, G. et al, International Agency for Research on Cancer, 1991

'Occurrence of nitro and non-nitro benzenoid musk compounds in human adipose tissue', Mauller, S. et al, *Chemosphere*, 1996; 33: 17–28

'Patch testing with fragrances: results of a multicentre of the European environmental and contact dermatitis research group with 48 frequently used constituents of perfumes', Frosch, P.J. et al, *Contact Dermatitis*, 1995; 33: 333–42

'Penetration of the fragrance compounds, cinnamaldehyde and cinnamyl alcohol, through human skin in vitro', Weibel, H. et al, *Contact Dermatitis*, 1996; 34: 423–6

'Pharmaceuticals and personal care products in the environment: agents of subtle change', Daughton, C. and Ternes, T., *Environmental Health Perspectives*, 1999; 107(suppl 6): 907–38

'Placebo controlled challenges with perfume in patients with asthmalike symptoms', Millquist, E. et al, *Allergy*, 1996; 51: 434–9

'Potential of carcinogenic effects of hair dyes', Shafer, N. and Shafer R.W., *NY State Journal of Medicine*, 1976; 76: 394–6

'Prospective study of talc use and ovarian cancer', Gertig, D.M. et al, *Journal of the National Cancer Institute*, 2000; 92: 249–52

'Relationship of hair dye use, benign breast disease, and breast cancer', Nasca, P.C. et al, *Journal of the National Cancer Institute*, 1980; 64: 23–8

'Solvents and neurotoxicity', White, R.F. and Proctor, S.P., *Lancet*, 1997; 349: 1239–43

'Some alkyl hydroxy benzoate preservatives (parabens) are estrogenic', Routledge, E.J. et al, *Toxicol Applied Pharmacology*, 1998; 153: 12–19

'Synergistic activation of estrogen receptors with' combinations of environmental chemicals', Arnold, S.F. et al, *Science*, 1996; 272: 1489–92

'The association between aluminum-containing products and Alzheimer's disease', Graves, A.B. et al, *Journal of Clinical Epidemiology*, 1990; 43: 35–44

'The estrogenic activity of phthalate esters in vitro', Harris, C. et al, *Environmental Health Perspectives*, 1997; 105: 802–11

'The relationship between perineal cosmetic talc usage and ovarian talc particle burden', Heller, D.S. et al, *American Journal of Obstetrics and Gynecology*, 1996; 74: 1507–10

'Toxic alert: A survey of "high street" companies and their approach to suspect chemicals', Friends of the Earth, October 2000.

'Toxic emissions from carpets', Anderson, R.C., *Journal of Nutritional and Environmental Medicine*, 1995; 5: 375–86

'Use of hair-colouring products and the risk of lymphoma, multiple myeloma and chronic lymphocytic leukemia', Zahm, A. et al, *American Journal of Public Health*, 1992; 82: 990–7

Books

Ashford, N. and C. Miller, *Chemical exposures: low levels and high stakes* (New York: Van Nostrand Reinhold, 1990)

Bremner, M., *Enquire within*, 3rd ed. (London: Helicon, 1994)

Cox, P. and P. Brusseau, *Secret ingredients* (London: Bantam, 1997)

Epstein, S., D. Steinman and S. LeVert, *The breast cancer prevention program* (New York: Macmillan, 1997)

Gosselin, R.E. et al, *Clinical toxicology of commercial products*, 5th ed. (Baltimore: Williams & Wilkins, 1984)

Harte, J. et al, *Toxics A to Z – a guide to everyday pollution hazards* (Los Angeles: University of California Press, 1991)

Lijinsky, W., *Chemistry and biology of N-nitroso compounds* (New York: Cambridge University Press, 1990)

Logan, K., *Clean house, clean planet* (New York: Pocket Books, 1997)

Mumby, Dr K., *The complete guide to food allergies and environmental illness* (London: Thorsons, 1993)

Neal's Yard Remedies, *Make your own cosmetics* (London: Aurum Press, 1997)

Pagram, B., *Natural housekeeping* (London: Gaia, 1997)

Palma, R.J., *The complete guide to household chemicals* (New York: Prometheus Books, 1995)

Ross Lewis, G., *1001 chemicals in everyday products*, 2nd ed. (New York: Wiley-Interscience, 1999)

Steinman, D. and S. Epstein, *The safe shoppers bible – a consumer's guide to nontoxic household products* (New York: Macmillan, 1995)

Steinman, D. and R.M. Wisner, *Living healthy in a toxic world* (New York: Perigree, 1996)

Wormwood, V.A., *The fragrant pharmacy* (London: Macmillan, 1990)

Index

acetone, 35
advertising of products, 14, 16–17,
    18–28
  air fresheners, 139, 140
  'antibacterial', 25
  'biodegradable', 23–5
  'biological', 184
  'CFC-free', 26
  fragrances, 29, 35
  'hypoallergenic', 23
  laundry detergents, 184
  'mild', 22
  'natural', 21–2, 133, 145
  'not tested on animals', 27–8
  'pH balanced', 25–6
  toilet cleaners, 154
  toiletries, 25–6, 71, 74, 88, 106
aerosols, 26, 86, 91, 122, 140,
    141–2, 148
  effects on health, 51, 141–3,
    144, 179
aftershave, 10, 13, 36, 37, 38
AHAs (fruit acids), 18, 67–8, 99,
    101, 110, 115, 159
air fresheners, 139–45
  alternatives to, 143–5
  carcinogens in, 140–41
  effects on health, 51, 141–3,
    144
  ingredients, 36, 37, 38, 140–42
alkyl compounds, 167–8, 173, 183
allergies, 31, 40–43, 57, 140
aloe extract, 104, 133
aluminium, 64, 67, 83–4
Alzheimer's disease, 38, 84, 187
ammonia, 10, 125, 148, 149, 156,
    168, 169, 174, 179, 197
antibacterial products, 25, 63, 64,
    96, 97–8

antioxidants, 95, 99, 103, 115
antiperspirants, see deodorants
aromatherapy candles, 145
arthritis, 48
asthma, 5, 31, 49–50, 140, 141
autoimmune diseases, 42, 47–8, see
    also immune system

baby lotion, 13
baking soda, 157, 158, 166,
    171–2, 188, 195
bath cleaners, 153, 158–60
bath foams, see bubble baths
benzaldehyde, 35–6
benzene, 118, 182, 191
benzophenones, 131, 132
benzyl acetate, 36, 186
benzyl alcohol, 36, 186
bicarbonate of soda, 81–2, 151, 157
biodegradable products, 23–5
birth defects, 52–4, 88, 160, 162,
    176, 189, 197
bleach, 146, 149, 151, 155, 174,
    183, 188–9
  chemicals in, 36, 37, 38, 156
  safety, 148
body lotions and oils, 127–9,
    130–31
  carcinogens in, 127–8, 129
  ingredients, 127–9
  plant extracts in, 129
body sprays, 88–9
body washes, 67–70
  alternatives to, 68–70
  carcinogens in, 67
  ingredients, 10, 67–8
borax, 157, 160, 165, 166, 169,
    170, 176, 188
breath sprays, 77

bubble baths, 64–7
  alternatives to, 68–70
  carcinogens in, 65, 66
  herbal, 65–7
  ingredients, 10, 13, 65–7
  and skin irritation, 64–5

caffeine, 41
camphor, 36, 186
cancer, 5, 15, 48–9
  breast, 44, 123–5
  calculating probability of, 6–7,
    91
  and chemical combinations,
    13–14, 17, 44–5
carbolic acid, 148–9, 156, see also
  cresol; phenol
carbomer, 73, 120, 128
carcinogens, 3, 4, 6–7, 10–13,
  48–9, see also under
  individual products
  and chemical combinations,
    13–14, 17, 44–5
  in fragrances, 33, 36, 37
  in hair dyes, 123–5
  in preservatives, 49, 66, 78,
    100, 103, 109, 115
carpet cleaners, 161–6
  alternatives to, 164–5, 166
  carcinogens in, 162, 163, 164,
    165
  deodorisers, 165
  ingredients, 161–4
  spot removers, 163–4
castor oil, 78, 85
children
  and bubble baths, 65
  exposure to toxins, 33–4, 142,
    161, 163, 168
  and fluoride toothpaste, 8,
    75–6, 80
  hyperactivity, 34
  and mouthwashes, 78
  and sun creams, 131, 134
  and talcum powder, 89–90
chlorides, 125
chlorine, 148, 156, 171,
  188–9

cleaners, all-purpose, 146–52
  alternatives to, 150–52
  carcinogens in, 147, 149, 150
  cream, 149
  ingredients, 147–9
  liquid, 146–9
  powder, 149, 150
  spray, 149
coal tar
  as a carcinogen, 112, 185
  dye, 66, 67, 149, 174
collagen, 129
cologne, 35, 36, 38
colourings, 23, 27, 49
  in bubble baths, 66
  in cleaners, 149, 169, 174, 179
  in facial cleansers, 94, 96
  in hair products, 119, 121, 125
  in mouthwashes, 78
  in soap, 63–4
  in toothpaste, 73, 77
consumer protection, 8–9, 16, see
  also regulations in manufacture
cornstarch, 86, 87, 90, 177
cosmetics, talc-based, 90
cresol, 141, 148, 156
cruelty-free products, 27–8

DEA (diethanolamine), 10–11, 12,
  65, 100, 106, 107, 109, 167,
  168
dental floss, 79
deodorants, 83–9
  alternatives to, 85–7
  carcinogens in, 85
  feminine, 91–2
  foot, 87
  and headaches, 51
  ingredients, 10, 35, 36, 37, 38,
    83–5
depilatories, 135
detergents
  biodegradable, 24
  cationic, 154–5
  dishwasher, 167–9
  ingredients, 35, 45, 102, 147,
    162, 173, 182
  manufacture of, 62, 182

detergents *(cont.)*
  in shampoos, 106–7, 109
  and skin reactions, 22
  and soap, 61–4
  vegetable-based, 22, 150
detoxification, 56–8
diabetes, 5, 48
diet, 49, 57
diseases, and toxic exposure, 5–8,
  9, 40, 45–56
dishwashing detergents, 167–72
  alternatives to, 169–70, 171–2
  carcinogens in, 167, 168, 169
  ingredients, 35, 36, 37, 167–9,
    170–71
disinfectants, 37, 146, 148, 149, 156
  natural, 158
drain cleaners, 193–5
dry cleaning agents, 189–92
  alternatives to, 191–2
  carcinogens in, 190–91

eating disorders, 55–6
EDTA (ethylene diamine tetra
  acetic acid), 147
electromagnetic pollution, 5
emollients, 114, 115, 128, 130
emulsifiers, 99, 102, 109, 114, 128
endocrine-disrupters 43–5, 52, 95,
  128
enzymes, 184
Epsom salts, 69–70
essential oils, 69, 81–2, 88–9, 90,
  123, 145, 151, 157
estrogen, 44, 48, 168
ethanol, 36, 118
ethoxylated alcohols, 12–13
ethyl acetate, 36–7, 186–7
exfoliating scrubs, 102

fabric softeners, 185–8
  alternatives to, 187–8
  ingredients, 36, 37, 38, 186–7
facial cleansers, 93–8
  alternatives to, 97–8
  antibacterial, 96, 97–8
  carcinogens in, 95
  ingredients, 10, 13, 93–7

fatigue, chronic fatigue syndrome,
  46–7
feminine freshness, 91–2
flavourings, 27, 29, 73, 77, 79
floor cleaners, 173–7
  alternatives to, 176–7
  carcinogens in, 173, 174
  ingredients, 173–6
  polishes, 175–7
flower waters, 88
fluoride
  and autoimmune diseases, 47
  in mouthwashes, 78
  in toothpaste, 7, 8, 74–6, 80
  in water, 5
food
  chemicals in, 4, 8, 11, 27, 29,
    40, 49, 53
  organic, 53, 57
foot deodorants, 87
formaldehydes, 4–5, 49, 50, 67,
  95, 140–41, 162, 163, 189
formulas in products, 17–18
fragrances, *see also* perfumes
  chemicals in, 10, 29–30, 32, 33,
    34, 35–8, 67
  effects on health, 31–9, 50, 51,
    142, 149, 162, 164, 184
  fragrance-free products, 23
  'natural', 18, 145
  and safety testing, 27, 30
  similarity to glue sniffing, 34–5
  and skin reactions, 31, 50,
    169
  in toiletries, 63, 85, 96
fruit acids (AHAs), 18, 67–8, 99,
  101, 110, 115, 159
furniture cleaners, 162–6
  fabric fresheners, 165–6
furniture polishes, 196–9
  alternatives to, 198–9
  ingredients, 196–7

glycerine
  in alternative products, 81–2,
    87
  in toiletries, 63, 64, 68, 95, 96,
    97, 99, 110, 115, 133

hair conditioners, 113–17, *see also* shampoos
  alternatives to, 116–17
  ingredients, 45, 114–16
  plant extracts in, 121
hair detanglers, 11
hair dyes, 10, 123–6
  alternatives to, 125–6
  and cancer, 123–5
hair gels, 119–21
  alternatives to, 121–3
  ingredients, 119–21
hair removers, 135
hair sprays, 117–19
  alternatives to, 121–3
  and headaches, 51
  ingredients, 10, 13, 35, 36, 37, 38, 45, 117–19
  plastic in, 117–18
hand creams, 10, 37, 130
hazardous waste list, 4, 33, 35, 36, 37, 186
headaches, 50–51
herb extracts, 18, 70, 126
homes, pollution in, 3–5, 142
hormone-disrupters, 3, 10, 43–5, 52, 53, 55, 65, 85, 117
hydrochloric acid, 155
hypoallergenic products, 23

immune system, 39, 41–2, 46, 49, 52, 54–5, *see also* autoimmune diseases
infertility, 3, 5, 52–4, 168

labelling, 6, 11, 15, 24, 27
  of perfumes, 31, 32
  of toothpaste, 74–5
lacquers, *see* varnishes
lanolin, 96
laundry detergents, 182–5
  alternatives to, 187–8, 191–2
  'biological', 184
  boosters, 188–9
  carcinogens in, 182–3, 184, 185, 186, 189, 190–91
  dry cleaning, 189–92
  fabric softeners, 185–7

  ingredients, 36, 37, 45, 182–3
  liquid, 185
  stain removers, 189, 192
laureth compounds, 107, 109, 121
limescale removers, 159
limonene, 37, 187
linalool, 37, 187
lindane, 45
lye, 17, 26, 96, 148, 155–6, 174, 193, 194, *see also* sodium hydroxide

make-up
  chemicals in, 10
  remover, 98
metal polishes, 199–201
methylene chloride, 37
mildew cleaners, 153, 155, 160
mineral baths, 69–70
mineral oil, 127, 130–31
moisturisers, 13, 94, 98–101
  alternatives to, 101
  carcinogens in, 99, 100
  ingredients, 98–101
mouthwashes, 50, 77–9
  alternatives to, 82
  ingredients, 77–9
multiple chemical sensitivity, 32, 187
multiple sclerosis, 38, 187

nail creams, 130
nail polish, 36, 37
  remover, 35, 36, 37
naphtha, 191, 196–7, 203
naphthalene, 64, 67, 155, 164, 199
'natural' products, 18, 21–2, 67–8, 133, 145
NDELA (N-nitrosodiethanolamine), 11–12, 107–8
nitro musks, 34
nitrosamines, 11–12, 107–8, 109, 128, 131, 147
non-Hodgkin's lymphoma, 124

optical brighteners, 162, 183
oven cleaners, 193, 194–5
  alternatives to, 195
  ingredients, 10

oxalic acid, 155, 164, 199
'ozone-friendly' products, 26

paint, 52
    remover, 36, 37
paraffin oil, 127, 130
parasites, 5, 54–5
Parkinson's disease, 38, 141, 187
PCBs (polychlorinated biphenyls),
    45
PEG (polyethylene glycol), 10,
    12, 100, 103, 107, 116, 168,
    184
perfumes, 32, 35, 36, 37, 38, *see
    also* fragrances
peroxide, 77
petroleum distillates, 168, 173,
    175, 189, 196, 197, 199
PG (propylene glycol), 10, 12,
    120–21, 128, 130
pH balanced products, 25–6, 107,
    208
phenol, 148–9, 156, 160, 167–8,
    197
phosphates, 146, 171, 174, 183
phosphoric acid, 159
phthalates, 45, 117
pine oil, 148, 149, 156, 173, 174,
    199
plants, as air purifiers, 58, 144
plastic
    chemicals in, 4, 45, 49
    in hair sprays, 117–18
poisoning, low-level, 41–3, 44
polishes
    car, 203
    floor, 175–7
    furniture, 51, 196–9
    marble, 202
    metal, 199–201
    shoe, 202–3
polysorbates, 12–13
potpourris, 145
pregnancy, and aerosols, 51
preservatives, 27, 94, 110,
    114–16
    carcinogens in, 49, 66, 78, 100,
        103, 109, 115

regulations in manufacture, 8–9,
    12, 23, 24, 29, *see also*
    consumer protection
reproductive problems, 3, 5, 52–4,
    168

saccharin, 73, 77, 78
safety testing, 8–9, 12, 13–14, 30,
    32, 38
    on animals, 12, 37, 44, 139,
        163
    GRAS (Generally Recognised as
        Safe), 18, 26–7, 51
scouring powder, 149, 150, 151
shampoos, 106–13, *see also* hair
    conditioners
    alternatives to, 111–13
    carcinogens in, 11–12, 107–8,
        109
    conditioning agents, 109–10
    dandruff, 112
    ingredients, 10, 13, 36, 45, 50,
        106–11
shaving cream, 102–5
    alternatives to, 104–5
    carcinogens in, 103, 104
    ingredients, 36, 37, 103–4
shoe polish, 202–3
shower cleaners, 153, 158–60
shower gels, 67–70
    alternatives to, 68–70
    carcinogens in, 67
    ingredients, 67–8
silica, 72, 150
skin reactions, 14–15, 22, 23
    from bubble bath, 64–5
    from facial cleansers, 94
    from fragrances, 31, 50
skin toners, 102
SLS (sodium lauryl sulfate), 10,
    22, 66, 72, 80, 106–7
smell, understanding sense of,
    30–31, 33, 34
soap, 61–4
    antibacterial, 63, 64
    carcinogens in, 63–4
    castile, 62, 68–9, 97, 113, 150,
        157, 170

soap (*cont.*)
  comparison with detergents, 61
  flakes, 170, 188
  genuine, 17, 61, 62, 68–9, 97
  ingredients, 10, 13, 36, 38, 50,
    63–4
  liquid, 64
  pH balanced, 107
sodium bisulphate, 141, 155
sodium chloride, 64, 66, 94, 110
sodium hydroxide, 17, 26, 96,
  182, 193, *see also* lye
solvents, 10, 34–5, 51, 54, 141
  as carcinogens, 33, 49, 54
  in cleaners, 146, 147–8, 162,
    174, 178, 199
  organic, 52, 178
  in polishes, 197
sorbitol, 77, 78
starch, 189
stearic acid, 64, 94, 128
sudden infant death syndrome, 38,
  187
sulphuric acid, 182, 193, 199
sun creams, 13, 131–5
  alternatives to, 134–5
  carcinogens in, 131, 132–3
  ingredients, 132–4
  'natural', 133
  and ultraviolet rays, 131–2
surfactants, 62, 67, 93–4, 106–7,
  147, 167, 209
synergism, 44–5, 209

talcum powder, 89–91
  alternatives to, 90–91
  and babies, 89–90
  in condoms, 89–91
  and ovarian cancer, 89
toilet cleaners, 153–7
  alternatives to, 156–7
  carcinogens in, 155
  ingredients, 154–6

toiletries, *see also under individual
    products*
  carcinogens in, 10–13
  chemicals in, 9–13
  shelf life of, 11, 18
toluene, 33, 191
toothpaste, 71–7
  alternatives to, 80–82
  carcinogens in, 72, 73
  fluoride in, 7, 8, 74–6, 80
  ingredients, 10, 50, 72–3
  whitening, 76–7, 82
TSP (trisodium phosphate), 151–2

upholstery cleaners, 162–6

varnishes, 4, 140, 155, 164
  remover, 36, 37
vaseline lotion, 36, 37, 38
vegetable oils, 95, 130, 131

washing powder, *see* laundry
    detergents
washing soda, 151, 170, 176, 183,
  188
washing-up liquid, *see* dishwashing
    detergents
water
  fluoridation, 5
  softener, 184
  toxins in, 5
window-cleaning agents, 178–81
  alternatives to, 180–81
  ingredients, 178–9
windshield wiper solution, 181
witch hazel, 86, 87, 98, 102
women
  chronic fatigue syndrome, 47
  and contraceptives, 89–91
  exposure to toxins, 33, 34, 142
  feminine freshness, 91–2
  pregnancy and aerosols, 51
  and talc-based cosmetics, 90